by Robert G. Skrentner

INSTRUMENTATION

HANDBOOK FOR
WATER & WASTEWATER
TREATMENT PLANTS

LEWIS PUBLISHERS

Library of Congress Cataloging-in-Publication Data

Skrentner, Robert G.
  Instrumentation handbook for water and wastewater treatment plants.

  Bibliography: p.
  Includes index.
  1. Water—Purification—Equipment and supplies—
Handbooks, manuals, etc. 2. Sewage—Purification—
Equipment and supplies—Handbooks, manuals, etc. I. Title.
TD433.S49   1987   628.1'62   87-31206
ISBN 0-87371-126-2

Third Printing 1989

Second Printing 1989

LEWIS PUBLISHERS, INC.
121 South Main Street, Chelsea, Michigan 48118

PRINTED IN THE UNITED STATES OF AMERICA

Chapter 1

Analytical Measurement

Analytical instruments perform on-line, real time analysis of process parameters that can also be analyzed in a laboratory. For water and wastewater applications, this includes chlorine residual, dissolved oxygen, pH, suspended solids, turbidity, respiration rate, total organic carbon, and various ion-specific analyses such as ammonia, nitrate, nitrite, etc.

This chapter includes amperometric chlorine residual, dissolved oxygen, pH, suspended solids, and turbidity.

TOTAL CHLORINE RESIDUAL

Applications

The most common method of disinfecting water and wastewater is chlorination. Free chlorine gas or hypochlorite acts as an agent to destroy microscopic organisms that are disease-producing or otherwise objectionable.

Chlorine residual analyzers measure the free or combined residual chlorine indirectly. Current technology is based on the assumption that maintaining a minimum chlorine residual (usually 1.0 mg/L) 30 minutes after adding chlorine will result in an effective disinfection level.

Principle of Operation

Several different measurement methods are used for chlorine residual, including colorimetric, amperometric, and polarographic. Amperometric is discussed here since it is the more common method measuring total chlorine residual.

The amperometric measurement method uses two dissimilar metals held in a solution or electrolyte. A voltage is applied to the two metals which act as electrodes. Electrons flow from the negative electrode to the positive electrode generating a current. Figure 1.1 illustrates the amperometric cell. The amount of current flowing between the electrodes is proportional to the amount of chlorine present in the solution.

The basic amperometric chlorine residual analyzer is illustrated in Figure 1.2. It consists of an inlet sample tank and flow regulator, reagent solutions with metering pumps, measurement cell, and electronic signal converter. The metered sample stream acts as the electrolyte as it flows through the measurement cell. Since chlorine in the sample can exist in many different chemical forms, the sample is conditioned with other chemicals in order for the cell to measure all chlorine present in the stream.

The current generated in the measurement cell is very sensitive to temperature variations. A reading can change as much as 3% per degree C temperature change. Therefore, automatic temperature compensation is necessary. A temperature sensor is located in the measurement cell to provide temperature feedback to the electronic converter. This feedback is then used to correct the indicator and output signals to compensate for the temperature effects.

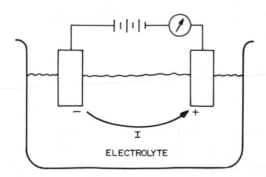

"I" PROPORTIONAL TO CI CONCENTRATION

Figure 1.1. Amperometric measurement.

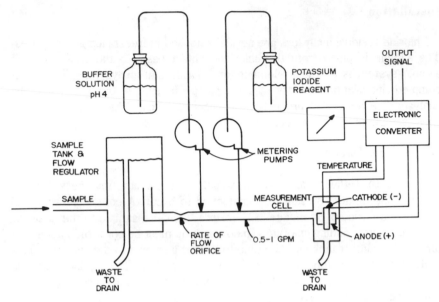

Figure 1.2.   Amperometric total chlorine residual analyzer.

Accuracy and Repeatability

Accuracy   ± 3% of full scale

Several ranges are normally available from 0–1 mg/L to 0–20 mg/L. The measurement error could range from 0.03–0.6 mg/L depending on the operating range used.

Repeatability   ± 1% of full scale

Automatic temperature compensation should enable this accuracy and repeatability to hold over a sample temperature range of 0–50°C (32–122°F).

Manufacturers' Options

Options available from manufacturers of chlorine residual analyzers are as follows:

1. Local indicator in the analyzer case
2. Supply of reagents
3. Integral solids filter
4. Output signal for remote monitoring of the chlorine residual

Installation

Chlorine residual analyzers are normally housed in freestanding enclosures. The sample is piped from the chlorine contact basin to the analyzer. The sample system is a critical element for a successful analyzer application. A complete installation consists of a sample point, sample transport, and the analyzer. Newer in-situ probe type analyzers eliminate the need for sample lines, sample pumps, and reagents.

Sample Point Location

Analyze the effluent after there is sufficient contact time between the chlorine and effluent stream for disinfection to occur. A commonly accepted disinfection period is 30 minutes. Therefore, deliver a sample to the analyzer 30 minutes after adding chlorine. To do this, you must be concerned about the time in the contact tank plus the time to deliver a sample to the analyzer. This is the total contact time as shown in Figure 1.3. A second 5 minute sample point or probe location is needed to implement automatic feedback control as discussed in Chapter 12.

Physically locate the sample point so it does not contribute unnecessary deadtime in chlorine residual analysis. In addition, take care to ensure the sample point is clean, thoroughly mixed, and representative of the monitored stream.

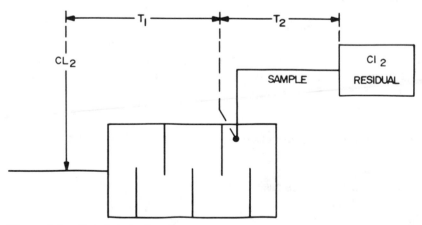

Figure 1.3. Sample point location.

Sample Transport

A sample line and pump are required to deliver the sample to the analyzer. Features of this sample transport assembly are shown in Figure 1.4.

- Select a pump capable of delivering 20–40 L/min (5–10 gpm).
- Size the pipe for a sample velocity of 1.5–3.0 m/s (5–10 ft/s).
- Determine the length of sample line so it will provide the desired transport time.
- Install a valve next to the analyzer so samples can be taken for calibration checks on the analyzer.
- Provide a source of clean water and required valves so the sample line can be backflushed to prevent plugging.
- If solids are present, install a filter.

Chlorine Analyzer

Install the analyzer so it is easy to service and maintain.

- Provide ample space, minimum of 1 m (3 ft), around all sides of the analyzer.

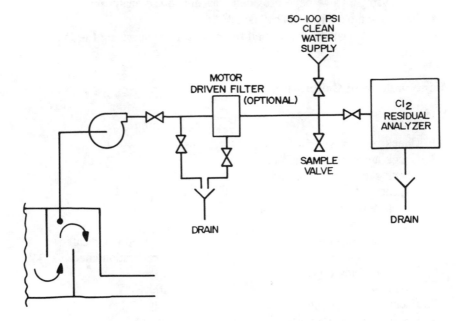

Figure 1.4. Sample transport.

- Locate the analyzer next to a floor drain.
- Provide a table nearby with the necessary equipment and chemicals to perform calibration checks.
- Provide a separate circuit on a lighting panel to power the analyzer.

Designer Checklist

If you can answer yes to the following questions when designing or reviewing chlorine residual analyzer applications, the application should be correct.

- Is remote monitoring required? If so, has an output signal compatible with the receiving instrument been specified?
- Is a local indicator provided?
- Is the sample point located so that during normal plant flows the sum of the contact tank time and sample transport time equals 5/30 minutes?
- Is the sample pipe length tuned to provide the required delivery time?
- Will the sample point location be thoroughly mixed and representative of the process stream?
- Are the sample pump and pipe sized to provide the recommended flow rates and velocities?
- Has a sample valve been provided adjacent to the analyzer?
- Can the sample line be backflushed?
- Is there adequate space around the analyzer for servicing the instrument?

Maintenance and Calibration

Task	Frequency
1. Check reagent supply	Daily
2. Check analyzer calibration	Daily
3. Check sample flow through analyzer	Daily
4. Check reagent flow to sample line	Daily
5. Calibrate analyzer	When need is indicated by calibration check
6. Replace tubing on reagent pumps	Monthly
7. Backflush sample line	Weekly
8. Clean analyzer drain lines	Weekly
9. Clean cell electrodes	Monthly

Deficiencies

The following problems have been encountered in existing chlorine residual analyzer installations:

- The sample point does not provide a representative mixed sample. This may be due to poor mixing, sample point location, or contact tank design.
- Contact time from the point of chlorine addition to the analyzer is too long.
- Sample lines plug and cannot be backflushed.
- No provision for taking a sample at the analyzer for calibration checks.
- Cramped space around the analyzer making maintenance difficult.
- No reagents in the analyzers.
- Using reagents with the wrong concentration.

DISSOLVED OXYGEN

Applications

Generally, dissolved oxygen (DO) meters in wastewater plants provide an approximate measurement of the oxygen available to support biological activity. In receiving waters, the DO meters monitor one parameter of water quality. DO meters are generally not used in water plants. (See Table 1.1 for DO meter application guidelines).

Principles of Operation

Dissolved oxygen meters consist of an electrochemical cell, the probe, and a signal conditioner or transmitter.

The two principal types of electrochemical cell used in DO probes are the galvanic cell and the polarographic cell. Galvanic and polarographic cells have very similar operating principles. Both cells consist of an electrolyte and two electrodes as shown in Figure 1.5.

Table 1.1. Dissolved Oxygen Meter Application Guidelines

Recommended	Not Recommended
Aeration tank	Chlorine contact tank
Oxygenation basins	H_2S bearing streams
Mixed liquor streams	
Secondary effluent	
Plant effluent	
Sample systems	

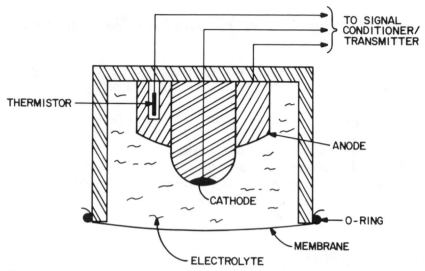

Figure 1.5. Membrane type DO probe.

The polarographic cell requires a polarizing voltage in the range of 0.5 to 1.0 Vdc. In one proprietary probe design, the electrodes are emersed directly in the process stream where process fluid acts as the electrolyte. However, in most commercially available probes, the electrolyte is contained by a gas permeable membrane, and oxygen is brought into contact with the electrodes by the action of diffusion. In either case, fluid flow past the probe greater than 30 cm/s (1 ft/s), is generally required to maintain a representative sample. Whether the probe is of the membrane or nonmembrane type, oxygen is reduced at the cathode, where the half cell reduction reaction is:

$$O_2 + 2H_2O + 4 \text{ electrons} \rightarrow 4 \text{ OH}^-$$

and at the anode, the anode metal is oxidized. The result of this oxidation/reduction process is a flow of electrons from the cathode to the anode proportional to the oxygen dissolved in the process stream.

The rate of this oxidation/reduction process is strongly affected by temperature. Therefore, accurate temperature measurement and compensation is essential to accurate DO measurement. Temperature is usually monitored by a thermistor located in the probe, and compensation is made in the signal conditioner/transmitter electronics.

Suspended and dissolved substances in the process stream can also affect electron flow. When solids accumulate on the membrane, they reduce the rate of oxygen transfer to the electrodes. In one design, a mechanical grindstone

Contents

Robert G. Skrentner is the Quality Manager for EMA Services Inc., a St. Paul-based engineering firm specializing in the application of instrumentation and computers to process control. EMA provides services in all phases of process control from conception and planning, through design and implementation, to training and staffing assistance. As Quality Manager, Skrentner is responsible for instrument and control system design standards and specifications, as well as maintaining information on instrument and control system experiences throughout EMA.

Prior to becoming Quality Manager, Skrentner was project manager for various study, design, and implementation projects in both water and wastewater applications.

Before joining EMA, Skrentner worked for the Detroit Water and Sewerage Department. While assigned to the Engineering Division, he worked on various instrumentation and control projects for the wastewater collection system, water distribution system, and wastewater treatment plant. Upon transferring to the Operations Division, he was placed in charge of the process control computer system at Detroit's 1200 mgd wastewater treatment plant.

Skrentner holds a BS and an MS in Civil Engineering from Wayne State University. He has contributed to a book on process control computer systems and one on combined sewerage monitoring and remote control, and has written numerous technical papers in the area of computer utilization in process control and control system management.

and maintainable lie outside manufacturers' manuals. Too often this knowledge is not shared outside the treatment plant because the persons responsible do not consider their solutions unique or important.

Although it was not a company project, I would like to acknowledge the continued support and encouragement of EMA Services, Inc. in producing this work. I would like to thank Bob Manross, the editor of the handbook upon which this book is based. Significant contributions to the handbook were made by Dag Knudsen, Rich Lackman, Claude Williams, and Walt Schuk.

This book is dedicated to those plant personnel who use and maintain instrumentation. They are hard-working and creative individuals who have overcome many obstacles to develop fixes and procedures for improved instrumentation performance.

Preface

Instrumentation is the group of devices used directly or indirectly to measure or control a variable. The term includes primary elements (sensors), final control elements (control valves and pumps), and switches, pushbuttons, controllers, annunciators, and related devices used to manipulate the variable.

Instrumentation, required for proper operation of water and wastewater treatment plants, was observed in various states of working order during visits to treatment plants. Personnel at these sites often criticized the instrumentation for failure to meet reliability and usefulness expectations. However, the same instrumentation performed satisfactorily in other locations.

This book provides information to engineers and operators about the prerequisites for success with instrumentation. The material in this book is intended as a guide for the selection, application, and maintenance of primary elements and final control elements.

The first six chapters cover basic, proven primary elements that meet specific needs and provide tangible benefits. For each instrument, the following information is provided:

1. Application
2. Principle of Operation
3. Accuracy and Repeatability
4. Manufacturers' Options
5. Installation
6. Designer Checklist
7. Maintenance and Calibration
8. Deficiencies
9. References

Chapter 7 discusses quality assurance techniques to ensure the primary elements are accurate and reliable. Chapters 8 through 11 cover final control elements (pumps and valves) applications. The last chapter covers process control instrumentation.

The information should not be considered all-inclusive. It is a beginning for what really works in the field. Answers to what makes instrumentation reliable continuously polishes the surface of a nonmembrane probe to keep the electrodes clean. To maintain gas permeability, fouled membrane probes must be manually cleaned.

Certain dissolved gases interfere with DO measurement by either nonmembrane or membrane probes. Common gases to be avoided are chlorine, hydrogen sulfide, carbon dioxide, and sulfur dioxide. Chlorine will be read by the probe as oxygen; carbon dioxide can neutralize some electrolytes; and hydrogen sulfide and sulfur dioxide can poison some metals used for an anode.

Accuracy and Repeatability

Accuracy ± 1 to $\pm 3\%$ of full scale at the calibration temperature

Additional error of $\pm 1\%$ can be expected for each 5.0°C (9.0°F) of change from the calibration temperature. Even with temperature compensation, an additional error of 3 to 4% can be expected over an operating range of 0 to 50°C (32° to 122°F).

A combined DO measurement accuracy of ± 2 to $\pm 4\%$ of full scale can be achieved in most meters under the conditions generally encountered at wastewater treatment plants. For a meter with a range of 0 to 10 mg/L DO, the measurement uncertainty is 0.4–0.8 mg/L. The uncertainty would double (0.8–1.6 mg/L) for a meter with a range of 0–20 mg/L.

The calibration of most probes will change after initial installation, or reinstallation after probe repair. Stabilization time, depending on the manufacturer, ranges from a couple of hours to several days. Output readings must be stabilized before the accuracies above apply.

Manufacturers' Options

1. A remote calibration unit can be installed near the probe to permit calibration where the transmitter is not within 50 ft or is not located within sight of the probe.
2. Ranges can be switch selectable. Some of the more common range selections are as follows:
   a. 0–3 mg/L, 0–15 mg/L
   b. 0–5 mg/L, 0–10 mg/L, 0–20 mg/L
3. Transmitter output signals
   a. 4–20 mAdc
   b. 10–50 mAdc
   c. 0–5 Vdc
4. Input power
   a. 115 Vac, 60 Hz

       b. 220 Vac, 60 Hz

       c. 24 Vdc

  5. Transmitter enclosure

       a. Panel mounted, NEMA 1B – general purpose

       b. Surface mounted, NEMA 4 – watertight

  6. Probe mounting

       a. Handrail brackets

       b. Tank side wall brackets

       c. Probe holder available in lengths of 3–6 m (10–20 ft)

  7. Probe cable in lengths of 7.5–15 m (25–50 ft)

  8. Agitator or ultrasonic cleaner for use where fluid velocity across the probe is less than 30 cm/s (1 ft/s)

  9. High/low alarm outputs

 10. Junction box for terminating probe cable if the transmitter or remote calibration station is more than 3 m from the probe connection head.

Installation

In Open Tanks and Channels

In most cases, open tanks have a convenient guard rail on which to mount the probe and transmitter (or the junction box, where the transmitter is remote from the probe). If the guard rail is not conveniently located, bolt the probe holder brackets to the free board area of the tank walls. The probe holder must be rigidly supported, but it must also be readily removable for probe maintenance. See Figure 1.6 for a typical configuration. Tilt the probe at an angle away from the general component of process flow to prevent air bubbles and debris from accumulating on the membrane. Submerge probes 60–90 cm (2–3 ft) in an area having sufficient agitation, and where the location is representative of the process. Determine the final probe location through testing during startup.

In Closed Oxygenation Tanks

Probe holders can be inserted in the process stream through a flanged opening in the tank cover. The holder must be rigidly fixed to the tank cover by a flange or quick connect sample port cover that is removable for probe maintenance. Also, a stilling well can be installed to provide a gas seal and a lateral support for an extended probe length. Probe placement in the process fluid is the same as in open tanks.

Mount the transmitter or junction box on a stand near the probe.

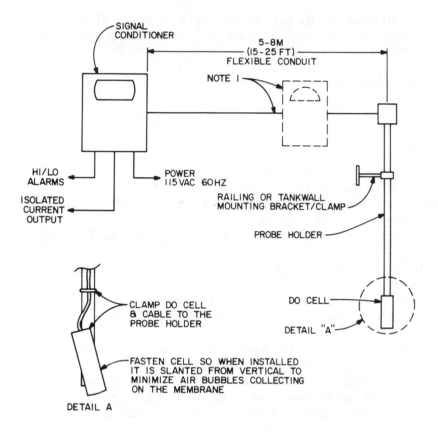

Figure 1.6. DO meter configuration.

On Pipelines

Do not subject probes mounted on pipelines to pressures greater than the manufacturer's recommendation, usually about 350 Pa (50 psi). The selected probe must be installed to ensure equal pressure on both sides of the membrane. Usually, probes are mounted in a tee in the pipeline with either a corporation seal on the probe or a bypass line to allow removal of the probe for maintenance. If the probe is part of a sample system, care must be taken to have a short transport time from the process to the probe. Where possible, direct measurement at the point of interest is preferable to transporting a sample for DO measurement.

Designer Checklist

If you can answer "yes" to the following questions when designing or reviewing dissolved oxygen meter applications, the application should be correct.

- If a nonmembrane probe is being considered, is the process stream's conductivity greater than 100 μmhos? Is the conductivity stable?
- Are interfering dissolved gases absent (chlorine, hydrogen sulfide, carbon dioxide, and sulfur dioxide)?
- Does the process fluid wash the probe at a rate greater than 30 cm/s (1 ft/s)?
- Does mounting eliminate the possibility of air bubbles being trapped at the measuring surface?
- Does the probe see a sample representative of the process?
- Is the probe mounted securely without causing likely collection sites for debris?
- Can the probe be removed easily for inspection or maintenance?
- Is installation of the meter designed so one person can calibrate it?
- Is the transmitter protected from the weather where necessary?
- Does the meter have automatic temperature compensation?
- Is the probe installed so it is always immersed in the process liquid?

Maintenance and Calibration

Membrane Probe Meters

Task	Frequency
1. Clean membrane	Depends on process stream characteristics. Some membranes must be cleaned daily; but more typically, every two days or once a week. Membrane breakage is a common problem. To reduce cleaning time, check calibration with a portable DO probe and clean membrane only when the two meter readings differ by more than 0.5 mg/L.

2. Replace membrane	Whenever membrane breaks or when electrolyte is replaced
3. Replace electrolyte	Every three to six months
4. Calibrate to portable probe	Every other day
5. Air calibrate	After membrane cleaning
6. Calibrate to standard	On initial installation or after major repair

Nonmembrane Probe Meters

Task	**Frequency**
1. Inspect and clean grindstone	Every two months
2. Replace grindstone	Every six months
3. Calibrate to standard	Weekly or biweekly are typical calibration intervals

Deficiencies

The following problems are commonly reported for dissolved oxygen meters.

- Agitator or cleaner becomes fouled with hair or fibers. Where possible, avoid use of agitators, mechanical cleaners, probe guards, or shields.
- The probe becomes fouled within a few hours due to process stream characteristics such as grease or slime growth.
- The probe cannot be withdrawn from the process stream because of mounting.
- One person cannot calibrate the probe because of the mounting bracket, difficult alignment, or awkward mounting.
- The probe cannot be calibrated by one person because the transmitter is too far from the probe and no remote calibration unit is installed.
- The probe is placed in a "dead" area of the tank. Poor mixing with the rest of the tank results in a false signal that does not show the true state of the process.

pH

Applications

Measurement of pH is one of the most important and frequently used tests in water chemistry. Practically every phase of water supply and wastewater treatment, e.g., acid-base neutralization, water softening, precipitation, coagulation, disinfection, and corrosion control, is pH-dependent. In addition,

Table 1.2.   Recommended Application of pH Sensors

Recommended	Not Recommended
Raw water	Digester sludge
Plant influent	
Primary effluent	
MLSS (if applicable)	
Plant effluent	

pH is used in alkalinity and carbon dioxide measurements and many other acid-base equilibria. Table 1.2 summarizes recommended applications for continuous monitoring with pH sensors.

Principle of Operation

All pH sensors employ a glass membrane electrode that develops an electrical potential varying with the pH of the process fluid. A reference electrode is used to measure the potential generated across the glass electrode.

Figure 1.7 shows a typical pH sensor arrangement. The heart of the sensor is the glass membrane. An electrical potential varying with pH is generated across the membrane. This potential is measured and amplified by an electronic signal conditioner. The complete electric circuit includes the glass electrode

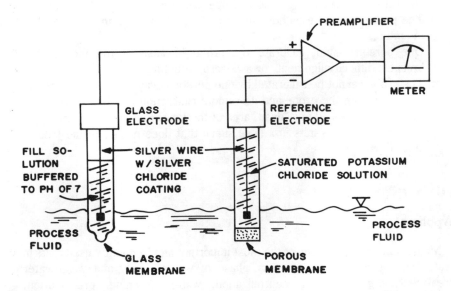

Figure 1.7.   Typical pH sensor.

wire, the glass membrane, the process fluid, the reference electrode fill solution and finally, the reference electrode wire.

Figure 1.8 shows an equivalent electric circuit of the pH sensor in Figure 1.7. Voltage at the input of the amplifier is:

$$E_i + E_r - E_g = 0$$
$$\text{and:}$$
$$E_i = E_g - E_r$$

Where E_i = amplifier input, mV
         E_g = glass electrode potential, mV
         E_r = reference electrode potential, mV

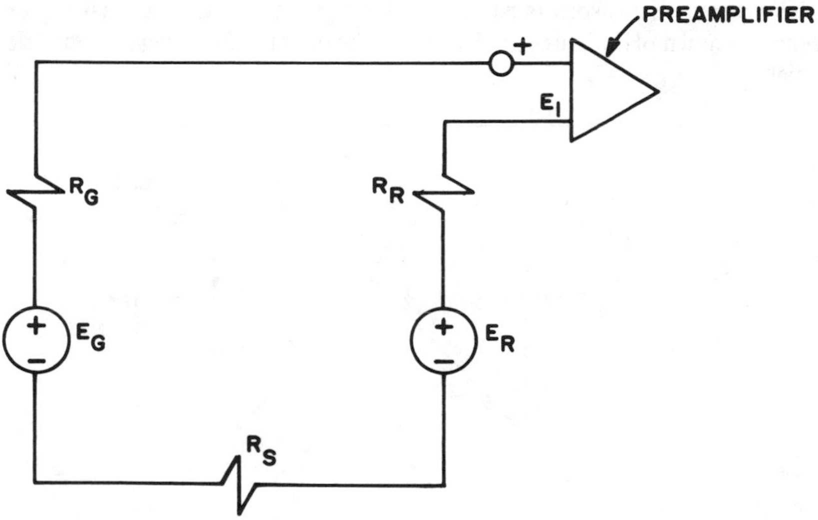

$$E = K_1 + K_2 (PH)$$

E_R = CONSTANT

R_G = RESISTANCE OF GLASS ELECTRODE

R_R = RESISTANCE OF REFERENCE ELECTRODE

R_S = RESISTANCE OF PROCESS FLUID SOLUTION

Figure 1.8.   Equivalent circuit.

The glass electrode has the approximate characteristic:

$$E_g = K_1 + K_2(pH)$$

Voltage at the amplifier input is:

$$E_i = K_1 + K_2(pH) - E_r$$

Where K_1 = asymmetric potential, mV
        K_2 = electrode gain, mV/pH

The reference electrode is designed so its potential E is constant with pH and other chemical characteristics of the process fluid. The asymmetric potential K_1 varies from sensor to sensor. It also changes as the sensor ages. For this reason pH sensors must be periodically standardized against buffer solutions of known pH. Figure 1.9 illustrates the effects of varying asymmetric potential.

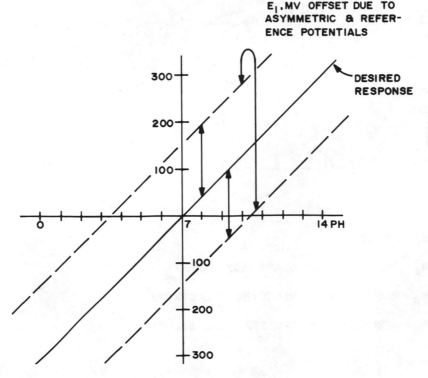

Figure 1.9.   Effect of varying asymmetric potential at constant (25°C) temperature.

The electrode gain K_2 is a function of temperature. For this reason most commercial pH sensors include automatic temperature compensation. A temperature sensor in the process fluid adjusts amplifier gain to compensate for changes in electrode gain which are caused by temperature. Figure 1.10 illustrates the effects of varying electrode gain.

Accuracy and Repeatability

Accuracy $\quad\quad$ ± 0.02 to ± 0.2 pH units. This represents the combined accuracy of the electrodes and the signal transmitter.

Most pH meters include automatic temperature compensation. Temperature effects are negligible with these meters. Without temperature compensation, an additional error of .002 pH per degree centigrade difference from the calibration temperature can be expected.

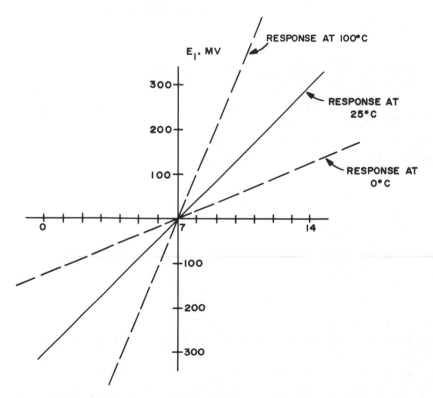

Figure 1.10. Effect of varying temperature at constant asymmetric potential.

Repeatability   ± 0.02 to ± 0.04 pH units
Stability   0.002 to 0.2 pH units per week

Stability (drift) is an important performance parameter that indicates how often meters must be recalibrated. With flow-through probe mounts, the velocity of the sample can cause a shift (0.2 to 0.3 pH) in measured values.

Methods of reporting performance specifications vary among manufacturers. Adjustment of the method of reporting performance specifications to equal units of measure will show there is a large variance in both the accuracy and the stability claimed by different manufacturers.

Manufacturers' Options

Most manufacturers sell several pH probes and mounting assembly configurations, each compatible with 3 or 4 transmitter/indicator/controller units called pH analyzers. Probe and analyzer options are listed below.

Probe Options

1. Mounting configurations for in-tank submersion, insertion in-process pipes, or side stream flow-through installation
2. Ultrasonic cleaning
3. Flow-powered cleaning
4. Mechanical wiper cleaning
5. Double-junction reference electrodes

Analyzer Options

1. Analog or digital (LCD) indicator
2. Signal outputs available:
   a. 10–50 mAdc
   b. 4–20 mAdc
   c. 0–1 Vdc
   d. 0–5 Vdc
3. Field selectable output spans from 2 to 14 pH in 2 pH increments
4. Alarm contacts
5. One manufacturer includes self-diagnosis for both electrodes and signal conditioner. A failure alarm contact output is enclosed.
6. Integrated process controller

Installation

The best installation of a pH meter, where pH control is not the objective of the measurements, is as part of a sample system along with other on-line analytical instruments. This locates the pH meter with other high maintenance instruments for ease of service. Buffer solutions needed for standardization can be conveniently stored with other analytical instrument reagents.

Use flow-through pH sensors in sample system installations. Provide bypass and shutoff valves for instrument removal and service. Select sensors with electrodes that are easily removed from the flow-through housing for cleaning and replacement. The flow-through sensor should be designed so that electrode tips are flush with the tube wall and do not obstruct flow. Locate the pH analyzer near the probe mounting assembly for easy standardization. Provide work surface for setting containers of buffer solution during standardization. Install a sample valve next to the sensor to collect a sample for conformance checks. Figure 1.11 shows sample system installation.

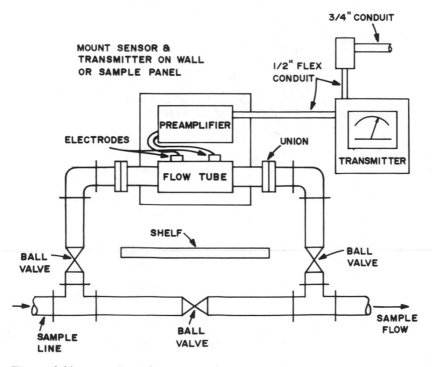

Figure 1.11.   Flow-through pH sensor installation.

For in-tank and open-channel installations, use a submersion type electrode assembly with an integral preamplifier. Figure 1.12 shows a typical installation. The electrode assembly is attached to a PVC pipe with a bracket normally mounted on a guardrail. Design the bracket so that the pipe and electrode assembly can be removed for maintenance without the use of tools. Secure all fastening devices to prevent dropping them in the tank or channel. Mount the signal conditioner/transmitter next to the electrode assembly mounting bracket. Provide enough spare cable to allow the sensor/pipe assembly to be lifted clear of the tank.

Install a submersion probe in a well-mixed zone at a point that will provide a representative sample of the process. If the probe is installed in an open channel, locate it in a free-flowing zone. Design the electrode assembly and support pipe installation to discourage collection of debris.

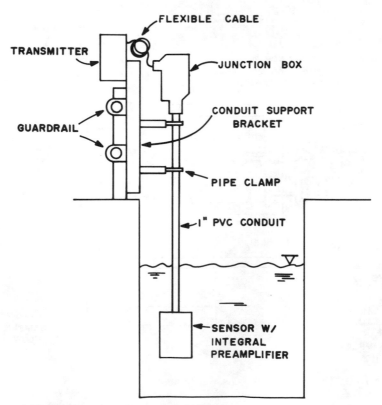

Figure 1.12.   Submersion pH sensor installation.

Designer Checklist

If you can answer "yes" to the following questions when designing or reviewing pH meter applications, the application should be correct.

- What are the process fluid temperature and pressure? Can the selected probe and probe assembly handle the expected range?
- Are all wetted parts PVC or stainless steel?
- Can the electrode assembly and the electrodes be removed easily for maintenance?
- Is the measuring system installation designed to allow maintenance and calibration by one person?
- Are the electrodes exposed to a representative sample of the process fluid?
- Is the electrode assembly securely mounted?
- Is there potential for debris to hang up on the electrode assembly?
- Is the process fluid likely to coat the electrodes?

Maintenance and Calibration

Task	Frequency
1. Clean electrodes	Depends upon process fluid. Monthly for plant effluent. Weekly for plant influent and sludges
2. Add reference electrode fill fluid	Weekly. Add as necessary (free-flowing type electrode)
3. Replace reference electrode	As dictated by operating experience (Nonflowing gel type electrode)
4. Standardization	Weekly after initial installation. Reduce to once per month if justified by experience
5. Transmitter calibration	Every six months

Deficiencies

The following problems are commonly reported for pH meters:

- Electrodes become coated with grease or sludge. Mechanical wipers have demonstrated some success in sewage treatment applications. Ultrasonic cleaners do not work on soft coatings like grease, oil, and sludges. The best solution to coating is periodic cleaning by trained personnel.

- Plugging of reference electrode. Switch to a double junction reference electrode when plugging is a problem.
- No provision for easy removal of probes for cleaning or replacement. Probes should be removable without shutting down process piping.
- The pH meter can't be easily calibrated or standardized because the transmitter/indicator is too far away from the probe.

SUSPENDED SOLIDS

Application

Suspended solids analyzers are used in wastewater treatment plants to continuously measure the concentration of solids in various process streams. Concentrations of interest range from effluent quality, 10–30 mg/L, to thickened sludge of several percent solids. In water plants, suspended solids are usually reported in terms of turbidity. A variety of instruments are commercially available to accommodate this wide spectrum of concentrations. This section will review light emitting and nuclear type solids analyzers. Table 1.3 presents application guidelines for suspended solids. Ultrasonic analyzers are not described in this chapter.

Principle of Operation

Suspended solids instruments are based on the attenuation or scattering of a beam of radiation. The type of beam used can be light, ultrasound, or nuclear.

Optical Technique

Optical techniques for measuring suspended solids are based on scattering of a beam of light by the suspended particles (see Figure 1.13). The portion of the light scattered is a function of the number and size of particles. Light transmitted through the stream is reduced in proportion to the light that is scattered. An instrument which can measure the scattered light, transmitted light, or both, provides a measure of the suspended solids present.

The optical type of suspended solids analyzer consists of a lamp which acts as a source of light and a photocell which measures the transmitted or scattered light (see Figure 1.14). Arrangement of the lamps and photocells depends on the manufacturer. An electronics package analyzes the received light and correlates this to the suspended solids in the sampled stream.

Since solids buildup and coating is a problem in wastewater plants,

Table 1.3.   Suspended Solids Application Guidelines

Recommended	Not Recommended
Optical Analyzers	
Solids concentrations from 20 mg/L–8%	Primary solids
Raw water	Flotation thickened sludge
Return activated sludge	Solids concentration greater than 8%
Waste activated sludge	
Mixed liquor	
Plant effluent/finished water	
Gravity thickened sludge	
Filter effluent	
Centrifuge supernatant	
Nuclear Analyzers	
Thickened sludge with concentrations greater than 8%	Streams with solids concentrations less than 15%
Centrifuged sludge	Streams with entrained air bubbles
	Line sizes larger than 35 cm (14 in.)
	Line size smaller than 15 cm (5 in.)
Ultrasonic Analyzers	
Solids concentration from 1–8% solids	Mixed liquor
Primary sludge	Secondary effluent
Waste activated sludge	Plant effluent
Return activated sludge	Pipe sizes greater than 30 cm (12 in.)
Gravity thickened sludge	Pipe sizes less than 10 cm (4 in.)

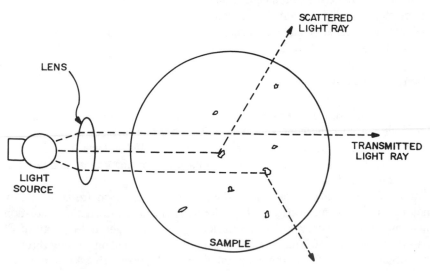

Figure 1.13.   Light scattering.

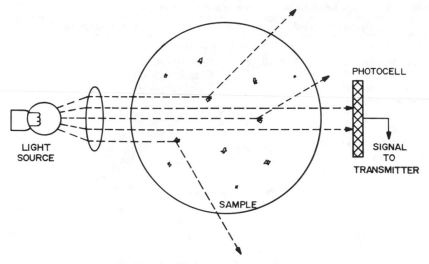

Figure 1.14. Transmissive type optical suspended solids analyzer.

manufacturers have devised several different methods to minimize or eliminate the effects of solids buildup. One technique uses a small sample chamber in which a piston draws and dispels a sample. The piston is designed with a flexible edge that mechanically cleans the glass which separates the lamp and photocell from the sample. Another design uses multiple lamps and photocells. By measuring the transmitted light at different angles to the lamp, comparisons can be made to nullify the effect of solids buildup. Still another design measures the light reflected at an angle to a falling stream; because the sample does not contact the lens, solids buildup does not occur.

Nuclear Radiation

A nuclear density gauge is a noncontact measurement of solids density. It does not measure percent solids directly; rather, it measures the specific gravity of the material. If the specific gravity of the liquid and solids is constant, then a correlation can be made between measured specific gravity and percent solids concentration.

In operation, a radioactive source emits gamma rays which are absorbed by material in the measured stream. High density materials absorb more radiation than low density materials. Thus a nuclear gauge averages the density of all material in the stream. The detector senses the total radiation passing through the stream to determine the material density. Figure 1.15 shows a solids analyzer.

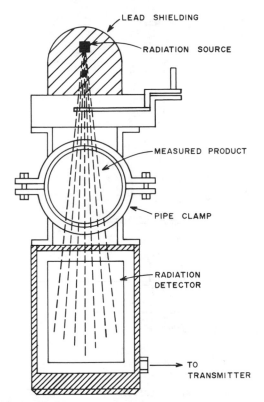

Figure 1.15.   Nuclear solids analyzer.

Turbidity

Turbidity can be classified as forward scatter or side scatter measurement types. In forward scatter turbidimeters, the measurement is in Jackson Turbidity Units (JTU). The JTU unit was derived from the Jackson candle turbidimeter shown in Figure 1.16. In this instrument, the sample is poured into the glass tube until the candle flame is seen to disappear, leaving a uniform field of light. At this transition point, the height of the column is read and converted into JTUs from a standard table.

Figure 1.17 shows a forward scattering type turbidimeter. This instrument measures the amount of light scattered by particles in the forward direction from the light beam. By establishing and maintaining a ratio of scattered light to the transmitted light, the effects of color changes can be eliminated and a direct measurement made of the particulates.

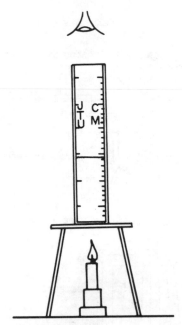

Figure 1.16. Jackson candle turbidimeter.

In side scatter turbidimeters, the turbidity is determined by measuring the amount of light scattered at some angle (usually 90°) from the light path by particles suspended in the sample. Figure 1.18 illustrates two styles of turbidimeters which use the side scatter method of measurement.

The units for side scatter turbidimeters are Nephelometric Turbidity Units (NTU). The word "nephelometric" describes the optical technique of measuring scattered light at an angle to the light path.

Formazin polymer has gained acceptance as the turbidity reference suspension standard. It is easy to prepare and is reproducible in its light-scattering properties. Although a sample of formazin suspension measured by forward scatter (JTUs) and side scatter (NTUs) turbidimeters will read approximately the same, they are not identical. A poor correlation exists when measuring a sample because of the variation in the absorption and optical scattering properties of the suspended particles. Because of this, turbidity units are not interchangeable between different types of turbidity meters. JTU or NTU can be correlated to suspended solids for a specific application.

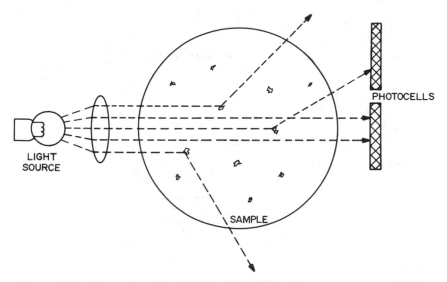

Figure 1.17.   Forward scatter turbidimeter.

Accuracy and Repeatability

Optical Solids Analyzers

Accuracy ±5% of full scale. Several ranges of operation are available. On a range of 0–3000 mg/L, the instrument error is ±150 mg/L of the actual reading. For a 0 to 10% range the error would be ±0.5%.

Repeatability ±1% of full scale

Nuclear Solids Analyzers

Accuracy ±0.5% of full scale. However, since the instrument measures specific gravity and is empirically calibrated to read out in percent solids the accuracy of solids measurement can be affected by changes in the specific gravity of the particulate or the fluid.

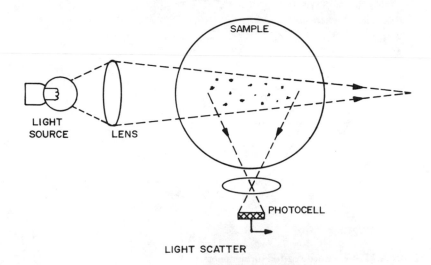

LIGHT SCATTER

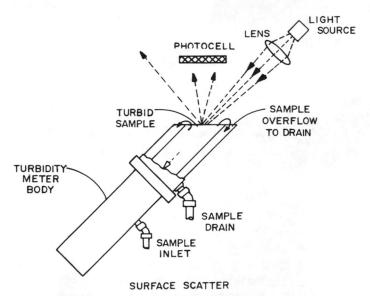

SURFACE SCATTER

Figure 1.18. Side scatter turbidimeters.

Turbidity Analyzers

Turbidity is a relative measurement. Accuracy and repeatability shown below is the capability of the instrument.

Accuracy	± 5% of full scale
Repeatability	± 2% of full scale

Manufacturers' Options

Some options are common to all types of solids analyzers. These common options are listed first followed by options which apply to a specific analyzer type.

Common Options

1. Low and high alarm contact outputs
2. Voltage or current output signals for remote monitoring. This is a standard offering on some analyzers.
3. Wall or panel mounting for the transmitter enclosure
4. Length of interconnecting cable between the sensor and transmitter

Optical Analyzers

1. Light shields to prevent stray light from introducing measurement errors
2. In-line pipe mounting adapters
3. Mounting brackets for installing on hand rails
4. Length or style (depends on manufacturer) of the sensor probe
5. Test standards for troubleshooting and calibration of transmitter electronics

Nuclear Analyzers

1. Mass flow computer (requires flowmeter input)
2. Pipe spool pieces with cleanout ports
3. Radiation source decay compensation
4. Automatic temperature compensation

Turbidity Analyzers

1. Sample system accessories such as pumps and bubble traps
2. Installation kits

3. Extended high ranges
4. Test standards for calibration

Installation

Installation details for solids analyzers are unique to each manufacturer. The variations of installation are too numerous to list here. Manufacturers installation manuals should be obtained and used when designing for a solids analyzer installation. Some general considerations for installing solids analyzers follow:

* Solids analyzers require frequent attention and calibration checks. Provide space for servicing and locate the sensor so it can be easily reached.
* If sample lines are required, make sure they are large enough and that flow velocity is high enough to minimize line plugging.
* Provide flushing water for the instrument and sample valve.
* Provide a sample valve next to the sensor so samples can be taken to check analyzer calibration.
* Mount the transmitter within sight of the sensor.
* Locate sensors or sample line taps where air bubbles are least likely to be present. Preferably a vertical line with an up-flow.

Designer Checklist

Refer to the manufacturer's installation instructions or operation and maintenance manuals for details of analyzer installation. Also review the general installation suggestions of this section. Caution: the performance of solids analyzers is directly related to ease of maintenance and calibration.

Maintenance and Calibration

Refer to the manufacturer's operation and maintenance manuals for specific recommendations on frequency of tasks. The following are general maintenance considerations for all suspended solids and turbidity analyzers.

Task	Frequency
1. Check analyzer calibration	Weekly
2. If sample line is used, check sample and drain flows	Daily
3. If sample line is used, backflush sample line	Weekly
4. Calibrate analyzer with a solution of known solids concentration	When need is indicated by a performance check

Deficiencies

Some problems encountered in existing solids analyzer installations are described below:

- The solids analyzer or the sample line tap is not located where there is a well-mixed representative process sample.
- There is no provision for taking a sample at the analyzer for calibration checks.
- Locating the sensor and/or transmitter for servicing and maintenance is difficult due to inaccessibility.
- Analyzer operating range does not match the range of solids in the process.
- For optical and turbidity analyzers, stray light causes erroneous readings.
- There are air bubbles in the sample.

BIBLIOGRAPHY

Total Chlorine Residual

1. Liptak, B.G., and K. Venczel. *Instrument Engineers Handbook of Process Measurement*, rev. ed. (Radnor, PA: Chilton Book Company, 1982).
2. "Manual on Installation of Refinery Instruments and Control Systems. Part II -Process Stream Analyzers," 3rd ed., API–RP550 (Washington, DC: American Petroleum Institute, 1977).
3. Kulin, G. "Recommended Practice for Measuring Residual Chlorine in Wastewater Treatment Plants with On-Line Analyzers." (Cincinnati, OH: U.S. Environmental Protection Agency).

Dissolved Oxygen

1. Liptak, B.G., and K. Venczel. *Instrument Engineers Handbook of Process Measurement*, rev. ed. (Radnor, PA: Chilton Book Company, 1982).
2. "Comparison of Field Testing of DO Analyzers." (Chicago, IL: APWA Research Foundation, September, 1982).
3. Kulin, G., and W.W. Schuk. "Evaluation of a Dissolved Oxygen Field Test Protocol." EPA 78-D-X0024-1 (Cincinnati OH: U.S. Environmental Protection Agency, 1978).
4. *Standard Methods For the Examination of Water and Wastewater*, 16th ed. (Washington, DC: American Public Health Association, 1985).

pH

1. Liptak, B.G., and K. Venczel. *Instrument Engineers Handbook of Process Measurement*, rev. ed. (Radnor PA: Chilton Book Company, 1982).
2. Considine, D.M., Ed. *Process Instruments and Controls Handbook*. (New York: McGraw-Hill Book Co., 1974).
3. Krigman, A. "Guide to Selecting pH and ORP Instrumentation," *InTech*, (August 1982), p. 31.

Suspended Solids

1. *Standard Methods for the Examination of Water and Wastewater*, 16th ed. (Washington, DC: American Public Health Association, 1985).
2. Condrashoff, G. *Wastewater In-Line Turbidity and Suspended Solids Measurements*. (Redwood City, CA: Monitor Technology, Inc.).
3. Simms, R.J. *A Return to Accurate Turbidity Measurement*. (Redwood City, CA.: Monitor Technology, Inc.).
4. Liptak, B.G., and K. Venczel. *Instrument Engineers Handbook of Process Measurement*, rev. ed. (Radnor, PA: Chilton Book Company, 1982).

Chapter 2

Flow Measurement, Closed Conduit Liquid Flow

There are many different types of flow meters. Each has an application for which it is best suited. Technologies used for flow meters include: magnetic, sonic and ultrasonic, mechanical (turbine and positive displacement), differential pressure, and vortex shedding. These technologies are used in three different categories of meters: full bore, insert, and clamp-on, depending on how the meter is mounted. Full bore meters are placed in the process piping and are not easily removed. Insertion meters are inserted into the pipe through a tap in the pipe wall. Clamp-on meters are affixed to the outside of the pipe.

Full bore meters include: orifice plates, venturis, magmeters, V-elements, turbines, positive displacement, vortex shedding, and target meters.

Insertion meters include: magnetic, target, some sonic and ultrasonic, pitot tubes, small turbines, and vortex shedding.

Clamp-on meters are sonic and ultrasonic, including dopplers.

This chapter includes magnetic, sonic and ultrasonic, turbines, venturi tubes and flow tubes, vortex shedding, and insertion magnetic meters. Orifice meters, pitot tubes, and turbines are covered in Chapter 3. Although these meters can be used for liquid flow, their use is limited. Orifice plates create high head losses and have limited rangeability. Pitot tubes have poor accuracy and repeatability and tend to clog. Other meters discussed in this chapter can be just as effective.

Positive displacement meters are widely used in water metering. They are highly accurate and have wide rangeability. They are totalizing type meters and are not suited to provide flow rate data.

Venturi tubes are also discussed in Chapter 3. Vortex shedding meters, discussed in this chapter, can also be used for gas flow but are not discussed in Chapter 3.

MAGNETIC

Applications

Magnetic flow meters (mag meters) are suitable for the application under the following general conditions:

1. Minimum head loss is desired.
2. The process fluid has a conductivity greater than 5 μmhos per centimeter.
3. The liquid is corrosive or abrasive.
4. The liquid has a solids concentration less than 10% by weight.
5. The pipe always flows full.

Mag meters are not recommended for the following applications:

1. Nonconducting liquid process streams.
2. Gas streams.
3. Streams with powdered or granular dry chemicals.
4. Liquid streams with a solids concentration greater than 10% by weight.

Application guidelines for mag meters are listed in Table 2.1. Older meters may have asbestos gasket material. Use care in handling these meters. Table 2.2 lists liner selection criteria for specific conditions.

Table 2.1   Mag Meter Applications Guidelines

Service	Liner Material	Gasket Material
Raw sewage	polyurethane	rubber, neoprene
Raw water	polyurethane, rubber	rubber, neoprene
Settled sewage	polyurethane	rubber, neoprene
Primary sludge	polyurethane or Teflon[a]	Teflon
Mixed liquor	polyurethane	rubber, neoprene
Return activated sludge	polyurethane	rubber, neoprene
Waste activated sludge	polyurethane	rubber, neoprene
Thickened sludge	polyurethane or Teflon	Teflon
Digester sludge	polyurethane or Teflon	Teflon
Digester supernatant	polyurethane or Teflon	Teflon
Polymer solutions	Teflon, rubber	Teflon
Clean (process) water	polyurethane, rubber	rubber, neoprene
Strongly corrosive	Teflon or Kynar	Teflon

[a]Registered trademark of E. I. du Pont de Nemours and Company, Inc., Wilmington, Delaware.

Table 2.2   Liner Selection Criteria for Specific Conditions

Liner Material	Resistance to Abrasion (Mild)	(Severe)	Resistance to Corrosion	Maximum Temperature
Teflon	good	not recom.	excellent	150°C (300°F)
Kynar	good	not recom.	excellent	100°C (212°F)
Polyurethane	excellent	excellent	not recom.	88°C (190°F)
Butyl rubber	excellent	good	not recom.	71°C (160°F)
Neoprene	excellent	good	not recom.	93°C (200°F)

Cost Considerations Regarding Liner Materials

If a base price is assumed for a polyurethane liner, the following costs may be used for comparison:

1. Rubber or neoprene costs will be approximately $5.90/cm ($15.00/in.) of meter diameter, greater than an equivalent meter with a polyurethane liner.
2. Teflon costs will be approximately $87.00/cm ($220.00/in.) of meter diameter, greater than an equivalent meter with a polyurethane liner.

Principle of Operation

Magnetic flow meters (mag meters) operate by using Faraday's principle of electromagnetic induction in which the induced voltage generated by an electrical conductor moving through a magnetic field is proportional to the conductor's velocity. Figures 2.1 and 2.2 illustrate the application of this principle to volumetric flow rate measurements.

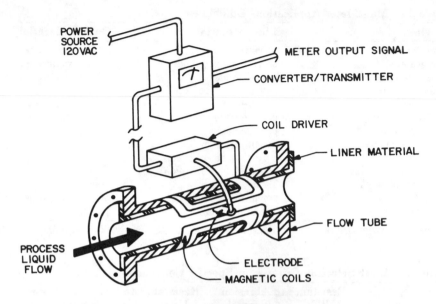

Figure 2.1.   Magnetic flow meter construction.

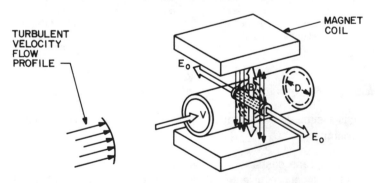

OUTPUT VOLTAGE = E_0 = KBDV x 10^{-4} VOLTS

WHERE K = COEFFICIENT TO ACCOUNT FOR NON-IDEALITY
B = MAGNETIC FLUX DENSITY, TELSA
D = PIPE DIA. (DISTANCE BETWEEN ELECTRODES), METERS
V = AREA - AVERAGE VELOCITY OF FLOW, M/S

Figure 2.2.   Mag meter induced voltage.

Commercial power is applied to the meter, and the coil driver energizes the magnetic coils which encase the spool pipe, creating a magnetic field. If the process liquid has enough conductivity, it will act as an electrical conductor and will induce an electrical voltage. This voltage is a summation of all the incremental voltages developed within each liquid particle occupying the magnetic field and is proportional to the field strength, pipe diameter, and "conductor velocity." The more rapid the rate of liquid flow, the greater the instantaneous value of electrode voltage.

The induced voltage is received by the two electrodes mounted 180° apart in the meter. This signal is sent to the converter/transmitter where it is summed, referenced, and converted from a magnetically induced voltage to the appropriate scaled output. The mag meter output is linearly proportional to flow.

Two basic types of mag meters are available, the ac mag meter and the dc mag meter. With the ac mag, line voltage is applied to the coils and a continuous flux is created producing a continuous low level ac electrode voltage. With the dc mag meter, the magnetic coils are periodically energized (pulsed), thereby producing two induced electrode voltages—one when energized, the other when deenergized. The energized electrode voltage is a combination of both true signal and noise, while the deenergized electrode voltage represents only noise. The difference between the two voltages is measured yielding a "clean" signal. Because of this operating scheme, the pulsed dc mag meters are zeroed every cycle whereas ac meters require stopping the flow for periodic rezeroing.

Dc mag meters are generally less bulky than ac mag meters and cost about 35% less. This is particularly true for the wafer style dc mag meters. The dc mag meter consumes less power than the ac mag meter and is somewhat easier to install and wire.

Where process flow is expected to change rapidly, the ac mag meter will follow the changes much better, since it operates at 60 Hz, while a dc mag meter operates from about 3 to 30 Hz. Ac mag meters handle slurries better because the hard particles in the slurry can create low frequency noise which affects the dc mag meters more than the ac mag meters.

Accuracy and Repeatability

Accuracy	± 1% of full scale
	± 3% of indicated flow when operating in the lower one-third of the meter range.
Repeatability	± 0.5% of full scale

These requirements can be met by the inherent characteristics of present-day mag meter design. However, the conditions described below will degrade these levels of operation.

Flow-disturbing piping obstructions located too near the meter inlet and outlet may add an additional 1–10% of uncertainty to the measured flow. Avoid locating the following obstructions nearer than five pipe diameters to the meter inlet or outlet:

1. valves, gates
2. tees, elbows
3. pumps
4. severe reducers and expanders (greater than 30° included angle)

Meter orientation leading to a nonfull meter pipe (i.e., trapped gases) or resulting in material buildup on the electrodes severely degrades accuracy.

Manufacturers' Options

Electrodes

1. Shape (see Figure 2.3)
2. Materials
   a. 316 stainless steel
   b. platinum/iridium
   c. tantalum
   d. Hastelloy
   e. nickel
3. Self-cleaning
   a. high frequency ultrasonic (continuous or portable)
   b. heat
   Most manufacturers state that the requirement for ultrasonic cleaning or removable electrodes is virtually eliminated with the high impedance electrode design. It is still an option with some manufacturers. If the process fluid can deposit mineral-like contaminants on the electrodes, ultrasonic cleaning may help to remove them. A process fluid

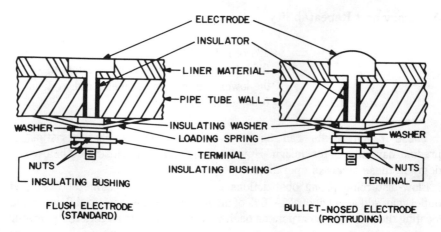

Figure 2.3. Mag meter electrode shape options.

with greasy contaminants probably cannot be cleaned ultrasonically and will require periodic electrode removal.

4. Field replaceable
Available with self-sealing liners only, e.g., neoprene or rubber (not available for Teflon-lined meters).

Mag/Flow Converter

1. Auto zero calibration
2. Output signal: 4–20 mAdc, digital pulse, scaled digital pulse
3. Face-mounted indicating meter

Miscellaneous Options

1. Grounding rings, straps, probes
2. The corrosive or abrasive characteristics of the process liquid dictate proper selection of the liner material and electrode construction (see Tables 2.1 and 2.2).
3. Corrosive resistant epoxy paint
4. Protected from accidental or continuous submergence, NEMA 6 submersible, watertight

Installation

- Locate the meter on the discharge side of pumps and on the upstream side of throttling valves.
- Locate the meter in a straight run of pipe free of valves or fittings with a minimum of five diameters upstream and downstream length.
- The process conduit must flow full of liquid.
- Meter sizing is critical. Size the meter to provide a fluid velocity within the following ranges:
  a. Nonsolids-bearing liquids: 1–9 m/s (3–30 ft/s).
  b. Solids-bearing liquids: 1.5–7.5 m/s (5–25 ft/s).
  c. Abrasive solids-bearing liquids 1–2 m/s (3–6 ft/s).
  Appropriate reducers/expanders may be required to achieve recommended operating velocities.
  Use the flow that will exist at startup for meter sizing. DO NOT USE 20-YEAR FLOW ESTIMATES FOR METER SIZING.
- The meter must have self-cleaning electrodes, ultrasonic or heated, for all applications except where process water is equivalent to, or better than, secondary effluent quality.
- Install the mag meter so it can be taken out of service for calibration and/or maintenance without disrupting the associated process. Recom-

mended isolation and bypass piping configurations are shown in Figure 2.4.

When the meter is to measure process liquids containing solids, e.g., primary sludge, RAS, WAS, thickened sludge, or when continuous electrode cleaning is not used, install a cleanout tee as shown in Figure 2.4. The decision to install bypass piping is a value judgment involving consideration of many factors including:

a. pipe size
b. available space
c. the ability to shut down the line while maintaining
   process operation or shifting to a parallel process unit.

- Properly ground all mag meters using stainless steel grounding rings and grounding straps supplied by the meter manufacturer. The grounding rings should have an inside diameter one cm (1/4 in.) less than that

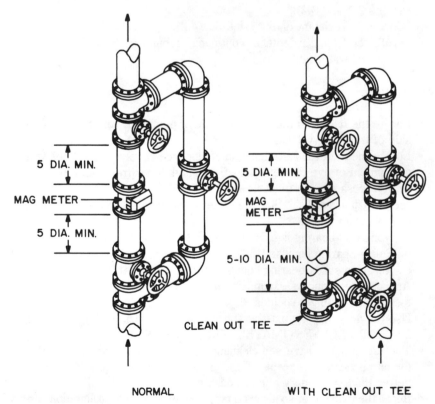

5 DIA. MIN.

MAG METER

5 DIA. MIN.

5 DIA. MIN.

MAG METER

5-10 DIA. MIN.

CLEAN OUT TEE

NORMAL   WITH CLEAN OUT TEE

Figure 2.4. Bypass pipe installation.

of the meter (for meters 10 cm [4 in.] in diameter or larger). Place them on both flanges with rounding straps as shown in Figure 2.5. Always ensure that the plant electrical system ground near the meter location provides adequate grounding. If a plant-wide grounding grid is available, ground the meter to it.

- Avoid locating mag meters near heavy induction equipment because it causes meter operational problems (100 hp motors and larger, no closer than 20 ft).
- Provide sufficient space to facilitate calibration, in-line maintenance, or meter removal.
- Orient the meter so the electrodes lie in a plane parallel to the floor.
- Wall-mount the transmitter/converter within sight of the meter in a NEMA 4 enclosure (NEMA 6 if possible submergence), or flush panel mounted so the cable length from the meter does not exceed 60 m (200 ft).
- Use driven-shield signal leads and route them between the transmitter and meter through dedicated 2 cm (3/4 in.) conduit. Route power wiring in separate conduit.
- Torque flange connections to the manufacturer's installation specifications.
- Wire power to the transmitter/converter and the coil driver through the same dedicated circuit. If separate circuits are used for transmitter/converter power and coil driver power, both circuits must originate from the same phase of the primary power feed.
- Mount the meter in a vertical pipe run with the flow direction upward. Install air bleed valves for meters mounted horizontally.

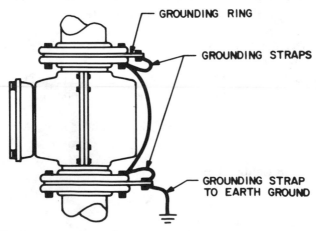

GROUNDING RING

GROUNDING STRAPS

GROUNDING STRAP
TO EARTH GROUND

Figure 2.5.   Mag meter grounding.

- Metered lines should not self-drain when shut down.
- Provide for flushing and filling with clean water in sludge applications where intermittent operation is expected.

Designer Checklist

If you can answer "yes" to the following questions when designing or reviewing mag meter applications, the application should be correct.

- Does the process liquid to be measured have a measured conductivity greater than 5 μmhos per centimeter?
- Will the pipe flow full under ALL conditions, excluding shutdown?
- If intermittent flow is expected, will the meter remain full at no flow? Does the transmitter have low flow zero cut-out circuitry?
- Does the meter size ensure a flow velocity between 1.5–7.5 m/s (5–25 ft/s) for solids-bearing liquids or 1–9 m/s (3–30 ft/s), for nonsolids-bearing liquids?
- Has the proper liner material been selected for the particular application?
- Do the electrodes require and have continuous cleaning capabilities?
- Are all piping elements/obstructions located at a minimum distance of five pipe diameters upstream or downstream of the meter?
- Have grounding rings and straps been provided, and is the meter grounded to a true ground?
- Have bypass piping and valving been provided?
- Is signal wiring between the transmitter and meter as specified by the meter manufacturer and routed in separate conduit?
- Has a dedicated power source been provided for the mag meter?
- Has the proper electrode material been selected to avoid excessive wear?
- Is the selected liner material compatible with the expected operating temperature?
- Does the design provide environmental temperatures within the range specified by the manufacturer for both the meter tube and the transmitter/converter?
- Are you using an ac mag meter for slurries or for flows that change rapidly?

Maintenance and Calibration

Task	Frequency
1. Calibrate transmitter	Monthly
2. Flow calibrate the meter	Every three months

Deficiencies

The following problems are commonly encountered with mag meter installations.

- Velocity skewing created by piping obstructions located too near to the meter cause accuracy problems and liner wear. In severe cases the obstructions cause the liner to be ripped away.
- Meter sizing does not maintain adequate flow velocities. Often this results from over-design for "future" flow rates.
- Improper installation results in nonfull pipe during low flows.
- Solids coating the electrodes due to lack of automatic electrode cleaning cause low flow velocity or intermittent flow.
- Meter or transmitter located so that calibration and maintenance accessibility are difficult.
- Isolation and bypass piping is not installed, requiring shutting down the process for meter zeroing and meter removal (when required).
- Infrequent calibration.
- No provisions are made for meter calibration.
- Improper grounding.

SONIC

Applications

Sonic flow meters are available in two basic types; the transmissive (through beam) type and the reflective (frequency-shift) or Doppler type.

The primary feature of sonic meters is repeatability, not accuracy.

Transmissive type (through-beam) sonic flow meters are suitable for the application under the following conditions:

1. Minimum head loss is desired.
2. The pipe always flows full.
3. The amount of suspended solids and entrained air bubbles in the process liquid together are "equivalent" to no greater than 0.3% suspended solids by weight.
4. Process liquid temperatures range between 0 and 80°C (32–180°F).
5. Line size is small enough so the sonic signal attentuation does not cause a problem. Consult meter manufacturer about applications in lines larger than 100 cm (42 in.).

Application guidelines for transmissive sonic flow meters are listed in Table 2.3. Reflective type (Doppler) sonic flow meters are suitable for the application under the following conditions:

Table 2.3   Transmissive Sonic Flow Meter Application Guidelines

Recommended	Not Recommended
Raw water	Raw sewage
Primary effluent	Primary sludge
Mixed liquor	Thickened sludge
Secondary clarifier effluent	Return activated sludge
Plant final effluent	Waste activated sludge
Finished water	
Process (wash) water	

1. Minimum head loss is desired.
2. The pipe always flows full.
3. The amount of solids and the entrained air bubbles in the process liquid must be equivalent to a suspended solids concentration greater than 25 mg/L but less than 4% by weight.
4. Flow velocities at the transducer must be maintained between 1–9 m/s (3–30 ft/s).
5. Pipe wall thickness must be less than 5 cm (2 in.).
6. The pipe is not constructed of, or lined with, an aggregate material.
7. The thickness of the pipe wall is exactly known.
8. A portable measurement device is desired.
9. Penetrating the process piping is to be avoided.
10. Low cost is a prime consideration.

Application guidelines for reflective sonic flow meters can be found in Table 2.4. A note of caution for the "recommended/not recommended" process applications. Determine the range of conditions under which a sonic flow meter will have to operate. Fluctuating flow conditions may cause intermittent operation of the meter.

Table 2.4   Reflective Sonic Flow Meter Application Guidelines

Recommended	Not Recommended
Raw water	Raw sewage
Primary sludge	Secondary clarifier effluent
Thickened sludge	Plant final effluent
Return activated sludge	Process (wash) water
Waste activated sludge	Finished water

Principle of Operation

Transmissive Sonic Meters

The transmissive sonic flow meter (also called through beam or time-of-travel meter) measures fluid velocity by measuring the difference in the time required for a sonic pulse to travel a specific distance through the fluid in the same general direction as fluid flow, and the time required for a sonic pulse to travel the same distance in the opposite direction. This meter is available in two types: (1) a pipe section with integral well-mounted transducers and (2) a direct-mounted version with the transducers mounted externally to an existing pipe. Both types use the same operating principle. Figure 2.6 shows the pipe section type.

Sonic transducers are energized alternately by electrical pulses and emit sonic pulses across the flow. The pulse whose directional component is downstream traverses the pipe in a shorter time than the pulse traveling against the flow (upstream). This time difference is proportional to the flow velocity, and an output signal linearly proportional to the flow rate is computed in the meter transmitter.

Reflective (Doppler) Sonic Meters

Figure 2.7 illustrates that operation of the reflective, or Doppler, sonic flow meter is based on a principle different from the transmissive type. The single

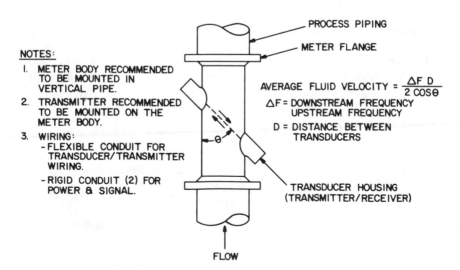

Figure 2.6.   Transmissive sonic flow meter.

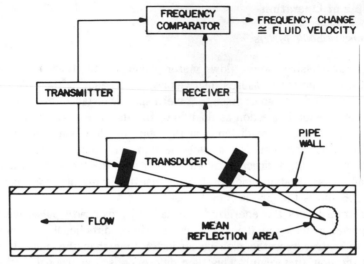

Figure 2.7. Reflective sonic flow meter.

transducer used is mounted on the external wall of the pipe. A signal of known frequency is sent into the fluid where it is reflected back to the transducer by suspended particulates or gas bubbles. Because the reflective matter is moving with the process stream, the frequency of the sonic energy waves is shifted as it is reflected. The magnitude of the frequency shift is proportional to the particle (flow) velocity and is converted electronically to the meter output signal linear to flow.

Accuracy and Repeatability

The accuracy and repeatability of sonic flow meters vary between the two types. The transmissive type provides a more accurate flow rate signal than does the reflective type. The following limits should be required when considering either type for any of the applications previously listed:

Transmissive:
(through beam)
    Accuracy   ±2% or less of actual flow wetted head type
                           ±5% or less for clamp-on type
    Repeatability   ±1% of actual flow

Reflective:
(Doppler type)
    Accuracy   ±5% of actual flow
    Repeatability   ±1% of actual flow

Sonic flow meters which meet the above accuracy may not meet the limits in operation. Several factors described below can degrade accuracy and must be adequately addressed during the design.

Flow-disturbing piping obstructions located too near the meter inlet may add up to 10% error to the measured flow rate. The following obstructions should not be located nearer than 7 to 10 pipe diameters from the meter inlet, or 5 pipe diameters from the outlet (on flow-tube, transmissive meters) or within these distances from the external transducer of the Doppler type meter:

1. valves (modulating and isolating) and gates
2. elbows and tees
3. pumps
4. severe reducers and expanders (greater than 30° included angle)

Skewing of the velocity profile will result if the recommended straight lengths of pipe are not provided upstream and downstream of the meter. Skewing will cause errors in the flow measurement.

Meter orientation leading to a nonfull pipe or resulting in material buildup or deposition will severely degrade meter accuracy.

The accuracy of Doppler meters is dependent on the velocity profile; amount, size, variation, and distribution of sound reflectors (solids or bubbles); line size; and flow meter design characteristics. If you can wet-calibrate the Doppler under actual operating conditions, $\pm 1\%$ accuracy can be obtained.

Manufacturers' Options

Transmissive (Through Beam) — Only

1. Meter tube construction:
   a. stainless steel
   b. carbon steel
2. Meter tube end connections:
   a. 150 lb ANSI RF flange
   b. 300 lb ANSI RF flange
   c. victaulic
   d. plain
3. Transducer mounting:
   a. wetted, with flush water port
   b. wetted, with epoxy window (Teflon-optional)
   c. wetted, removable without process disruption

Reflective (Doppler) – Only

Transducer mounting:

1. External, clamp-on
2. Wetted, with flow tube

Common Options

1. Input power:
   a. 115 Vac, 50–60 Hz
   b. 220 Vac, 50–60 Hz
   c. 24 Vdc
2. Transmitter:
   a. Outputs:
      1) 4–20 mAdc
      2) 0–10 Vdc
      3) pulse rate
   b. Integral flow rate indicator/totalizer
   c. Adjustable relay contact alarm outputs
3. Environment:
   Temperature: -20 to 60°C (-4 to 140°F).

Installation

- Install transmissive sonic flow meters having wetted transducers so the meters can be taken out of service for calibration or maintenance without disrupting the associated process. Recommended bypass configurations are the same as those recommended for magnetic flow meters and are shown in Figure 2.4. The decision to install bypass piping involves such considerations as pipe size, available space, and the ability to shut down the line while maintaining process operation or shifting to a parallel process unit.
- Locate meters on the discharge side of pumps and on the upstream side of throttling valves if these devices are near the required meter location.
- Flow velocities through the meter should be maintained between 1–9 m/s (3–30 ft/s). Transmissive meters can read down to 0.3 m/s (1 ft/s). Appropriate reducers/expanders may be required to achieve recommended operating velocities.
  USE THE FLOW THAT WILL EXIST AT START-UP FOR METER SIZING.

- Provide straight runs of pipe upstream and downstream of the meter as described under Accuracy and Repeatability.
- Orient spool-piece type meters in, or locate clamp-on type meters on, vertical process piping where possible, with flow direction upward only. Install air bleed valves when horizontally mounted.
- Metered lines should not self-drain when shut down.
- Locate the meter in an accessible location with sufficient space for calibration, in-line maintenance, or meter removal. Install the transmitter as close as possible (3.7 m, 12 ft) to the clamp-on transducer. When a meter tube is used, mount the transmitter directly on the tube.
- Install clamp-on transducers according to the manufacturer's suggested procedures. Be sure that no air bubbles are present in the epoxy sealant compound.
- Use separate conduit to wire line power and signal wiring.
- Follow precisely the manufacturer's guidelines for aligning sonic transducers to the pipe.
- Install the transmissive meter as a spool piece for pipes ranging in size from 91 to 7.6 cm (36 to 3 in.). For pipes smaller than this, mount the transducers in an axial configuration as shown in Figure 2.8.
- If the process piping is such that a clamp-on type meter is not practical, consider an insert type Doppler. This meter uses the Doppler principle

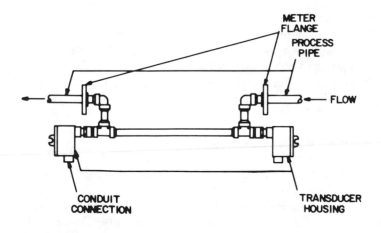

NOTE:

1. METER TUBE SHALL ALWAYS BE MOUNTED IN HORIZONTAL PROCESS PIPE.

Figure 2.8. Transmissive sonic meter, axial configuration.

but is inserted similar to an insertion mag meter as discussed later in this chapter.

Designer Checklist

If you can answer yes to the following questions when designing or reviewing sonic flow meter applications, the application should be correct.

Common Items

- Will the pipe flow full under all conditions?
- If intermittent flow is expected, will the pipe remain full at no-flow conditions?
- Are all piping elements/obstructions located a minimum distance of 7 to 10 pipe diameters upstream and 5 diameters downstream?
- Are the meter and transmitter easily accessible?
- Will adequate flow velocities be realized, 1–9 m/s (3–30 ft/s)?
- Is the meter located correctly in relation to pumps and throttling valves?
- Does the design provide environmental temperatures within the manufacturer's specified range?
- Has a sample of the process fluid been tested for sonic transmittance?

Transmissive (Through Beam) Items

- Have the proper spool piece material and end connections been provided for?
- Is the transmitter mounted on the spool piece?
- Is the process liquid recommended in Table 2.3? Is the amount of air bubbles and solids less than 3% by volume?
- Is the process liquid temperature between 0–80°C (32–180°F)?

Reflective (Doppler) Items

- Is the process liquid recommended in Table 2.4? Does it have a solids and/or air bubble content sufficient for the meter?
- Is the clamp-on transducer located where no excessive pipe or liquid-transmitted vibration will occur?
- Is the pipe inside diameter and wall thickness known accurately?
- Are the piping and tanks such that you can wet-calibrate the unit?

Maintenance and Calibration

Task	Frequency
1. Calibrate	Every two months

Deficiencies

The following problems are commonly encountered with existing sonic flow meter installations:

- Piping obstructions located too near to meter, causing accuracy problems.
- Meter sizing is such that adequate flow velocities are not maintained. Many times this results from over design for "future" flow rates.
- Installation resulting in nonfull pipe during low flows.
- Solids coating, due to low flow velocity or intermittent flow.
- Meter or transmitter located so that calibration and maintenance accessibility is difficult.
- Infrequent calibration.
- No provisions for flow rate testing/calibration.
- Solids concentration or entrained air greater than acceptable for transmissive type, resulting in poor accuracy or an unacceptable signal.
- Solids concentration or entrained air less than required for reflective type, resulting in poor accuracy or an unacceptable signal.
- Grease and scum buildup on pipe walls and wetted transducers.

TURBINE

Application

Turbine flow meters are suitable for the application under the following conditions:

1. The typical head loss through a turbine meter of 21–35 kPa (3–5 psi) can be tolerated.
2. The process pipe flows full under all conditions.
3. The process liquid is relatively "clear," i.e., a solids concentration less than 0.1% by weight (1000 mg/L) and is free of fibrous materials and/ or debris.
4. A maximum meter rangeability of 10:1 is acceptable.
5. An intermittent flow may be expected.
6. The process fluid's viscosity range is 2–15 μm^2/s (2–15 cSt).

Application guidelines for turbine flow meters are shown in Table 2.5.

Table 2.5   Turbine Flow Meter Application Guidelines

Recommended	Not Recommended
Raw water	Raw sewage
Plant final effluent	Primary sludge
Finished water	Secondary sludge (RAS & WAS)
Secondary clarifier	Mixed liquor
effluent	Primary effluent
Process (wash) water	Chemical slurries
Steam condensate	

Principle of Operation

Turbine flow meters consist of a pipe section with a multibladed impeller suspended in the fluid stream on a free running bearing as shown in Figure 2.9. The direction of rotation of the impeller is perpendicular to the flow direction, and the impeller blades sweep out nearly the full bore of the pipe. The impeller is driven by the process liquid impinging on the blades. Within the linear flow range of the meter, the impeller's angular velocity is directly proportional to the liquid velocity which is, in turn, proportional to the volumetric flow rate. The speed of rotation is monitored by an electromagnetic pickup coil which operates either on a reluctance or inductance principle to produce a pulse. The output signal is a continuous voltage pulse train with each pulse representing a discrete volume of liquid. Associated electronics units then convert and display volumetric flow (flow rate) and/or total accumulated flow.

Due to the nature of their linear-to-flow relationship, turbine meters must be

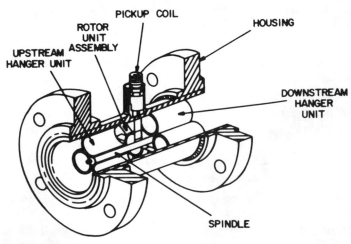

Figure 2.9. Turbine flow meter.

properly sized by volumetric flow rate. A meter sized to a specified range of linear flow rate measurement should not be used for flow rates outside that range.

Follow these guidelines when sizing a turbine meter:

1. The flow meter should be sized for 120–130% of the maximum expected process flow rate.
2. If the meter is sized by volumetric flow rate (guideline No. 1), it will have a diameter smaller than the process pipe. See Figure 2.10 for reducer/straight pipe installation.
3. If the meter size is the same diameter as the process pipe, its range will be severely reduced (to 2:1 or 3:1); however, the head loss through the meter will be less than if volumetrically sized.
4. Liquid cavitation may occur in the meter if upstream line pressure is not sufficient. To ensure sufficient pressure, the downstream line pressure must be a minimum of 2 times the meter head loss plus 1.25 times the liquid vapor pressure. If this condition cannot be met, a larger size meter with a correspondingly reduced meter range is required.
5. Available turbine meter sizes range from 0.5 to 60.0 cm (3/16 to 24 in.) in diameter.

Accuracy and Repeatability

The accuracy and repeatability characteristics of turbine flow meters, when properly applied and installed, should be:

Accuracy	$\pm 0.25\%$ of actual flow, within the linear range of the meter
Repeatability:	$\pm 0.05\%$ of actual flow, within the linear range of the meter

Each turbine flow meter has a unique "K" factor (the number of pulses per unit volume) which is determined during factory calibration. This factor is adversely affected by two conditions:

1. The liquid viscosity is significantly greater than that of clean water. This condition should not occur in water and wastewater applications recommended in Table 2.5.
2. The moving components become impaired by buildup of solids and/or fibrous materials.

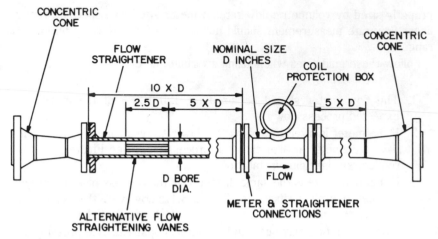

Figure 2.10.   Turbine meter mounting.

Manufacturers' Options

Turbine flow meter options available to the buyer are limited due to meter design standardization. Typically, options are limited to wetted parts materials for additional protection against corrosion and some additional equipment listed below:

1. Wetted parts materials
   a. stainless steel (standard)
   b. Hastelloy C
   c. P.T.F.E. (bearings)
2. Flow straightening vanes/elements
3. An additional electromagnetic pickup and associated electronics for increased accuracy
4. Turbine meters may require additional equipment for secondary read-out and/or transmitter devices. These should be purchased from the meter manufacturer. If another supplier is used, take care to ensure that both units are compatible with regard to pulse shape, amplitude, width, and signal frequency.
5. Typical secondary elements may include:
   a. electromechanical rate indicator and totalizer
   b. pulse-to-current transducer
   c. signal pulse preamplifier (for long distance pulse signal transmission)

Installation

- Flow-disturbing piping obstructions severely affect turbine meter accuracy. Figure 2.10 shows the recommended installation piping and details, including a flow-straightening element.
- When a flow-straightening element is used, the flow-disturbing effects of the following obstructions will be adequately damped in a minimum upstream distance of 10 pipe diameters (including the straightener). NOTE: If no flow-straightening element is used, extend this minimum distance to 25 to 30 pipe diameters.
  a. Valves and gates
  b. Tees and elbows
  c. Severe reducers and expanders (greater than 30 degrees included angle)
- Locate piping obstructions no nearer than 5 pipe diameters downstream from the meter.
- Install the meter in a horizontal pipe run.
- Install the meter on the discharge side of pumps and on the upstream side of throttling valves.
- Shield the cable between the turbine meter and electronics. Minimize cable length and do not route it through areas of high electrical noise.

Designer Checklist

If you can answer "yes" to the following questions when designing or reviewing turbine flow meter applications, the application should be correct.

- Is the intended process liquid as recommended in Table 2.5?
- Can the expected head loss be tolerated from a hydraulic standpoint?
- Is the expected upstream line pressure great enough to prevent cavitation in the meter?
- If the meter diameter is equal to the process pipe, is an instrument rangeability of 3:1 acceptable?
- Are all pipe obstructions located at least a distance of 5 pipe diameters downstream and a minimum upstream distance of:
  a. 10 pipe diameters (when a flow straightener is utilized)?
  b. 25–30 pipe diameters (when a flow straightener is not utilized)?
- Will the turbine meter be full under flowing conditions?
- Is the process liquid essentially free of solids, fibrous materials, or debris?
- Has the proper secondary flow indication device(s) been provided; has line power been provided, if required, for the device(s)?
- Is the meter easily accessible for maintenance?
- Have provisions been made for calibrating the meter?

Maintenance and Calibration

Turbine meters normally do not require periodic calibration. When accuracy becomes questionable (as observed through performance monitoring), examine it to determine maintenance requirements, or if none are required, determine a new "K" factor by hydraulic testing.

The meter constant "K", in clean water, is determined by the manufacturer prior to meter shipment. If the intended application process liquid has physical characteristics that significantly differ from clean water, consult the manufacturer for additional testing data.

Deficiencies

The following problems are commonly encountered with existing turbine meter applications:

* Inadequate upstream and downstream straight run piping, resulting in poor meter accuracy.
* No flow straightening vanes, resulting in poor accuracy.
* Meter sized too large, resulting in nonfull pipe and/or poor accuracy at low flows.
* Meter sized correctly, but reducers located too near inlet and outlet.
* Meter applied to a process liquid with an excessive solids concentration.

VENTURI TUBES AND FLOW TUBES

Applications

Recommended applications for venturi and proprietary flow tubes as primary elements include nearly all water and wastewater treatment process streams. The most critical aspect of proper application is the type of pressure sensing system used to measure the differential pressure produced by the primary tube.

Therefore, the recommended applications for these instruments are listed by type of pressure sensing system. The major types of sensing systems are:

1. Open connection, without flushing (including piezometric rings).
2. Open connection, with flushing (excluding piezometric rings).
3. Diaphragm sealed connections.

Table 2.6 presents the application guidelines for venturi and flow tubes. Additional requirements for venturi and flow tube applications are:

Table 2.6   Venturi and Flow Tube Application Guidelines

Recommended Without Flushing	Recommended With Flushing or Diaphragm Seals	Not Recommended
Secondary effluent	Raw sewage	Primary sludge
Final effluent	Primary effluent	Thickened sludge
Process (wash) water	Return activated sludge	Chemical (corrosive) slurries
Raw water	Waste activated sludge	
Finished water	Mixed liquor	

1. The metering tube must flow full.
2. A maximum to minimum measuring range of 4:1 is acceptable.
3. The Reynolds number of the process flow at the meter should be greater than 150,000.
4. Do not use venturi or flow tube meters in line with a positive displacement pump. The resultant flow pulsations will produce excessive signal noise and measurement inaccuracy.

Principle of Operation

Venturi Tube

A venturi tube operates on the principle that a fluid flowing through a pipe section that contains a constriction of known geometry will cause a pressure drop at the constriction area. The difference in pressure between the inlet and the constriction area (throat) is proportional to the square of the flow rate. Figure 2.11a shows a cutaway of a typical venturi tube.

General Equation

$$W = 353Yd^2[hr/(1-\beta^2)]^{1/2}$$

Where   d = throat diameter of pipe
D = pipe inside diameter
h = differential produced (in inches of water)
Y = net expansion factor
β = d/D (Beta ratio)
r = specific weight
w = flow rate (pounds/hr)

Several manufacturers provide differential-causing flow tubes which are modified versions of the classical venturi tube. These devices operate on the same principle as the classical venturi. They provide features which make them

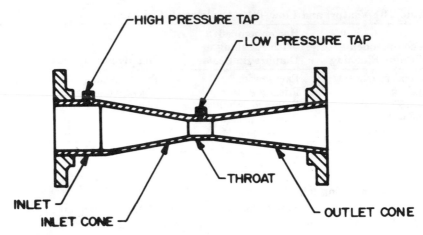

Figure 2.11.   Classic Herschel venturi.

more attractive for some applications, e.g., less space is required for installation, less overall head loss, and lower installed cost. Figure 2.11b shows three commonly used flow tubes.

At Reynolds numbers below 40,000, most differential head elements deviated significantly from the square root relationship between flow and differential pressure. The V-element meter creates a differential pressure that conforms to the square root relationship down to Reynolds numbers as low as 500. The meter can be used for viscous liquids and low flows. Figure 2.12 shows a V-element meter.

The V-shaped restriction of this element creates the differential pressure. The V element has no critical surface dimensions or sharp edges that must be kept within strict tolerances. Its upstream face is slanted to minimize damage from the impingement of solids.

The venturi tube, the proprietary flow tubes, and the V-element are primary sensing elements and require a secondary element to measure pressure differential.

The differential pressure created by a venturi or flow tube is generally measured by connecting a differential pressure (DP) transmitter to the sensing taps with pipe or tubing (tap lines). The discussion here focuses on three common DP transmitter connection methods. Further information regarding the DP transmitter and tap line runs is presented in Chapter 6.

Open Connection

For this method, the venturi tube pressure taps are connected directly to the DP transmitter and the tap lines are allowed to fill with process liquid. To avoid

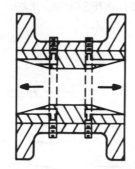

GENTILE OR BETHLEHEM
FLOW TUBE

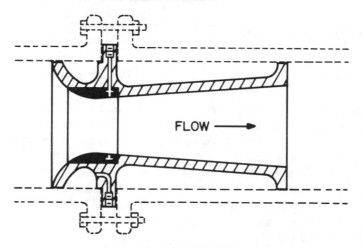

FLOW ⟶

LO-LOSS FLOW TUBE

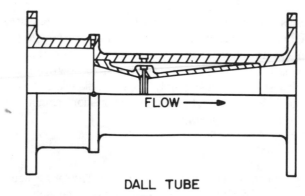

FLOW ⟶

DALL TUBE

Figure 2.11b. Proprietary flow tubes.

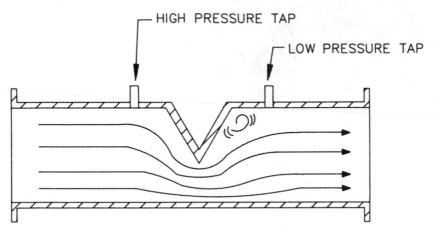

Figure 2.12. V-element meter.

tap line clogging, do not use this method for liquids with greater than 30 mg/L solids.

Providing a flushing water system for this sensing method allows the measurement of process liquids containing solids which would normally clog the sensing lines. There are two methods of operating flushing water systems. In one, flushing water is applied intermittently to purge solids from the tap line (measurement is interrupted during the purge cycle). In the second, a continuous equal flow of purge water is applied to both taps to act as a barrier to solids. As the purge water back pressure is measured in the latter method, it is critical that purge flows are equal. Direct and flushing water connections are illustrated in Figure 2.13.

Piezometric Rings

Piezometric rings may be used to sense inlet and throat pressures. These are normally used in very large diameter tubes where an average pressure is required to compensate for velocity profile variations. The rings consist of several holes for each tap (in a plane perpendicular to flow) connected to an annular ring. They should be used only on clean liquids. Flushing water systems cannot be used because the purge water short circuits within the annular ring to the nearest tap hole.

Diaphragm Sealed Sensors

The diaphragm sealed sensor method allows solids-bearing liquids to be measured without a tap flushing system. The process liquid is separated from the tap lines and transmitter by a diaphragm.

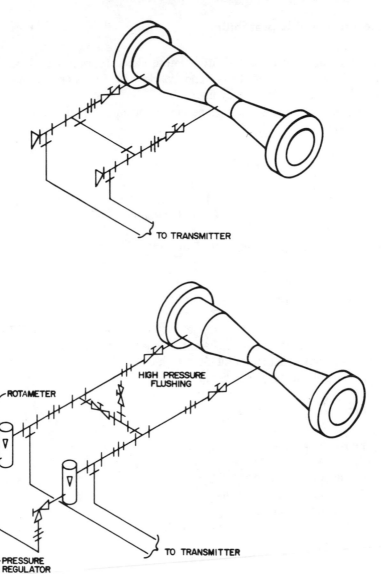

Figure 2.13.   Typical differential piping.

Accuracy and Repeatability

The accuracy and repeatability of venturi and proprietary flow meters vary. The characteristics of the secondary element (transmitter) must also be included in the total accuracy figure.

The following limits are generally attainable for the previously listed applications:

Classical Venturi and DP Transmitter

        Accuracy   ± 1% of actual flow
        Repeatability   ± 1% of actual flow

Proprietary Flow Tubes and DP Transmitter

        Accuracy   ± 1 to ± 3% of actual flow
        Repeatability   ± 1% of actual flow

These levels of accuracy reflect optimum values achievable for proper application and installation. Factors which degrade these levels include improper tube sizing (Beta ratio too high or low for the expected flow range), and piping elements which disrupt the velocity profile.

Manufacturers' Options

Primary Meter Tube

1. Single sensing ports
2. Piezometer ring sensing
3. Inspection openings
4. Manual rodders for cleanout of the sensing ports

Sensing System

1. Conventional
2. Diaphragm sealed

Secondary Element (Transmitter)

1. Differential pressure transmitter (see Chapter 6)
2. Manometer transmitter (not recommended)

Installation

Primary System

- Venturi and flow tubes may be installed in any position to suit the requirements of the application and piping as long as the meter flows full.
- For best accuracy, flow disturbing obstructions (fittings) should not be located too near the meter inlet. The following guidelines indicate the minimum upstream distance of straight pipe recommended between the fitting and the meter inlet:

a.	Reducers	8 diameters of reduced pipe size
b.	Expanders	4 diameters
c.	Fully open valve	5 diameters valve
d.	Check valve	12 diameters
e.	Throttled gate or ball valve	20 diameters
f.	90° bend(s)	4 diameters

For more information on lengths of straight pipe runs, see Bibliography numbers 8, 9, and 10.
- If a flow control valve is required in the line, it should be placed a minimum of 5 pipe diameters downstream of the meter tube. As seen in item e. above, to place the valve upstream requires a much greater straight run of pipe.
- Locate all downstream pipe fittings a minimum distance of 4 pipe diameters downstream of the throat tap(s).
- Place the meter tube a minimum distance of 10 diameters downstream from the pump discharge. The meter tube can be located on the suction side of a centrifugal pump only if subatmospheric pressure can be avoided.
- Install the primary and secondary flow elements in an accessible location with suitable space for maintenance and calibration.
- Orient single-tap meters (at inlet and throat) in the process piping so that the taps lie in the upper half of the meridian plane.

Secondary System

- Place the differential pressure transmitter below the hydraulic grade line to facilitate positive gas bleeding.
- Place an indicator gauge (DP) near the primary element for convenience in calibration and performance checking.

The following refer to all installations except those having diaphragm-sealed sensors (see Figure 2.13).

- If there is a possibility of the tap lines freezing, use insulation and heat tape to wrap them.
- Install tap line (connecting) tubing so that it has a minimum downward slope from the meter of 1:12.
- Install a bleed valve or gas collector at the highest point in the tap line run.
- Provide valves to isolate the transmitter for calibration.
- Provide a flushing water system if the process liquid contains greater than 30 mg/L of solids.
- Use connecting tubing no smaller than 1 cm (3/8 in.) in diameter.

The following refer to installations where continuous flushing is required (see Figure 2.13).

- The head loss in the tubing between the flushing water connection and the sensor tap should be the same in both lines so the pressure differential is unaffected.
- The flushing water supply pressure should be at least 70 kPa (10 psi) higher than process pressure.
- Equip the flushing water supply line for each tap with a rotameter for visual inspection and adjustment of purge flow.

Designer Checklist

If you can answer "yes" to the following questions when designing or reviewing venturi and proprietary flow tube applications, the application should be correct.

- Is the process liquid recommended in Table 2.6 compatible with the type of meter under consideration?
- Will the meter tube flow full?
- Is a maximum to minimum measurement range of 4:1 acceptable?
- Is the Reynolds number at the meter expected to be 150,000 or greater? If not, consider a V-element or target meter.
- Has the meter tube been sized to accommodate the present flow range (bear in mind that meters sized for a 20-year projected flow are typically oversized)?
- If the meter is to measure a solids-bearing liquid:
  a. Are single sensor taps being used as opposed to a piezometric ring?
  b. Has either a flushing system or diaphragm-sealed sensor system been provided?

- Is adequate straight run piping provided up and downstream from the meter tube?
- Have provisions been made for bleeding and flushing tap lines?
- Are the tap lines sloped properly?
- Are both the meter and secondary elements readily accessible?
- Will the meter be placed in a process line having smooth dynamics, e.g., not pulsating as in positive displacement pump applications?

Maintenance and Calibration

Primary System

Task	Frequency
1. Operate manual rodders (bayonets), if used, or flushing water (if used)	Weekly or when performance monitoring indicates decrease of differential pressure
2. If annular rings are included with the tube, bleed off gas	Weekly at first until history is established
3. Test the primary with a portable manometer	If performance monitoring indicates an accuracy change

Secondary Systems

Task	Frequency
1. Bleed tap lines of entrapped air	Weekly or first until history is established
2. Re-calibrate the transmitter using a portable manometer.	Monthly

Deficiencies

The following problems are commonly encountered with existing venturi and flow tube applications:

- Meter oversized, low flow measurements are lost due to square root function cutoff.
- Meters installed in process lines having pulsating flow (reciprocating pumps), causing erroneously high flow rates.
- Tap lines inadequately sloped and/or not provided with bleed valves, causing gas buildup.
- Improper differential range selection for the DP transmitter.
- Inadequately designed flushing systems which skew the pressure differential.

• Insufficient straight run piping upstream and downstream of the meter.

VORTEX SHEDDING

Applications

Vortex shedding flow meters are suitable for use under the following general conditions.

1. Medium head loss is acceptable.
2. The process fluid is a relatively clean liquid or gas.
3. The pipe always flows full.
4. The Reynolds number is greater than 10,000.
5. The process fluid does not contain air bubbles.
6. You want to replace an orifice meter.

The unit can also be used for clean gases and steam. Vortex shedding flow meter application guidelines can be found in Table 2.7.

Principle of Operation

The wetted parts of a vortex shedding flow meter consist of a bluff body and a vortex sensing element. The bluff body is an upstream obstacle producing vortices which are detected by the downstream sensing element.

During flow, a slow-moving boundary layer of fluid is formed on the outer surface of the bluff body. The sides of the bluff body shed this boundary layer alternately, forming low pressure eddies or vortices behind it as shown in Figure 2.14.

The frequency at which the vortices are shed is directly proportional to the fluid velocity. A flagpole is an example of a bluff body. Vortices shed by the pole cause the flag behind it to flap in the wind.

Ways to sense the vortices shed from the bluff body include the following:

Table 2.7   Vortex Shedding Flow Meter Applications Guidelines

Recommended	Not Recommended
Raw water	High viscous fluids
Secondary effluent	Mixed liquor
Finished water	Dirty liquids which can coat or wear the bluff body
	Raw sewage
	Sludges

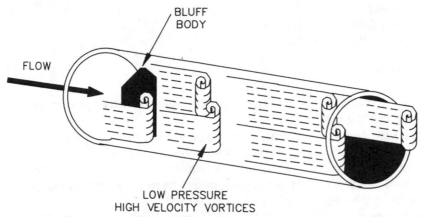

Figure 2.14. Vortices formed by bluff body.

1. A piezoelectric sensor inside the bluff body but outside the process fluid. It can sense the change in stress on the bluff body caused by the changing low and high pressure zones.
2. A dual bluff body arrangement with diaphragms imbedded on either side of the second bluff body that senses the pressure fluctuations (vortices). The diaphragm movement can be converted to either variable strain or variable capacitance readings.
3. An ultrasonic beam downstream from the bluff body is modulated by the vortices.

Accuracy and Repeatability

Accuracy	$\pm$ 1% of full scale liquids
	$\pm$ 1.5% of full scale gases
Repeatability	$\pm$ 0.5% of full scale
	$\pm$ 10:1 turndown for liquids
	$\pm$ 20:1 turndown for gases

Manufacturers' Options

1. Body
   a. flanged
   b. wafer
2. Transmitter
   a. two wire loop powered, or self contained
   b. output signal: 4 -20 mAdc, scaled digital pulse at 0–1 kHz or 0–10 kHz
3. Hot tap for bluff body replacement

4. Power
   a. 115 Vac, 50–60 Hz
   b. 230 Vac, 50–60 Hz

Installation

- Locate the meter on the discharge side of pumps and on the upstream side of throttling valves.
- Locate the meter in a straight run of pipe free of valves or fittings with a minimum of 15 diameters upstream to elbows and fitting, 50 diameters upstream to control valves, and five diameters downstream to obstructions. If a flow straightener is used, mount the meter 8 diameters downstream. See Figure 2.15 for additional details.
- The process conduit must flow full of liquid. Vertical pipes with liquid flowing up is ideal. In horizontal runs, ensure full pipe flow. See Figure 2.16 for meter orientation.
- Pipe sizing is critical. Size the pipe to provide a fluid velocity of 1–9 m/s (3–30 ft/s).
- Provide sufficient space to facilitate calibration, in-line maintenance, or meter removal.
- Metered lines should not self-drain when shut down.

The vortex shedding flow meter is similar to an orifice plate with a beta ratio of 0.7. See the orifice plate section in Chapter 3 for additional details.

Designer Checklist

If you can answer "yes" to the following questions when designing or reviewing mag meter applications, the application should be correct.

- Will the pipe flow full under ALL conditions, excluding shutdown?
- Are all piping elements/obstructions located at minimum pipe diameters upstream or downstream of the meter?
- Is signal wiring between the transmitter and meter as specified by the meter manufacturer and routed in separate conduit?
- Does the design provide environmental temperatures within the range specified by the manufacturer for both the meter tube and the transmitter/converter?
- If the process fluid is dirty, does the specified meter not have ports which can become clogged?
- If you expect high flows, does the specified meter not have moving parts which can wear under high flows?
- Is the fluid viscosity low?

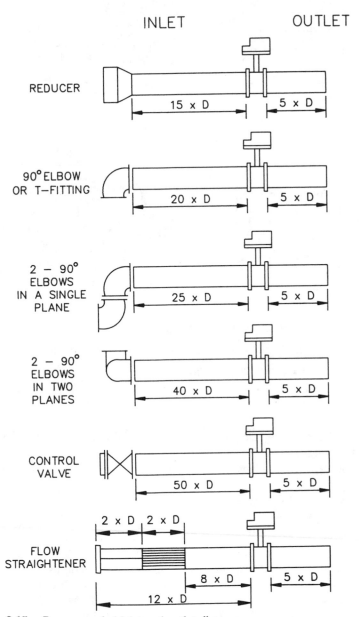

INLET   OUTLET

REDUCER
15 x D   5 x D

90° ELBOW
OR T-FITTING
20 x D   5 x D

2 - 90°
ELBOWS
IN A SINGLE
PLANE
25 x D   5 x D

2 - 90°
ELBOWS
IN TWO
PLANES
40 x D   5 x D

CONTROL
VALVE
50 x D   5 x D

2 x D 2 x D

FLOW
STRAIGHTENER
8 x D   5 x D
12 x D

Figure 2.15.   Recommended inlet and outlet distances.

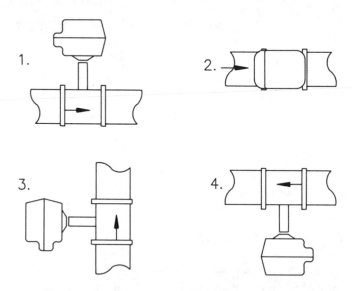

1 OR 2 — LIQUID APPLICATION (FULL PIPE)

2, 3, OR 4 — HIGH TEMPERATURE APPLICATIONS

1 — CRYOGENIC LIQUIDS

Figure 2.16.   Recommended meter orientation.

- If you have a noisy electrical environment, is the signal-to-noise ratio high enough?
- If you expect vibrations, have you specified a capacitive type sensor?

Maintenance and Calibration

Task	Frequency
1. Calibrate transmitter	Monthly
2. Flow calibrate the meter	Every three months

Deficiencies

The following problems could be encountered with vortex shedding meter installations:

- Velocity skewing created by piping obstructions located too near to the meter causes accuracy problems.

- Pipe sizing does not maintain adequate flow velocities. Often this results from over-design for "future" flow rates.
- Solids coating the bluff body due to low flow velocity or intermittent flow.
- Severe wear, corrosion, or buildup on the bluff body affects vortex frequency.
- Flow too low to provide proper sensitivity. Newer bluff body shapes produce stronger vortices.
- Meter or transmitter located so that calibration and maintenance accessibility are difficult.

INSERTION MAGNETIC

Applications

Operating conditions where insertion magnetic flow meters are suitable include:

1. Streams in which some additional head loss can occur.
2. Liquids with a conductivity greater than 5 μmhos per centimeter.
3. Liquid streams with a solids concentration less than 10% by weight.
4. The Reynolds number is greater than 4000.

Insertion magnetic flow meters are not recommended for the following applications:

1. Nonconducting liquid process streams.
2. Streams with powdered or granular dry chemicals or which contain abrasive materials.
3. Liquid streams with a solids concentration greater than 10% by weight.

Application guidelines for insertion magnetic flow meters are presented in Table 2.8.

Table 2.8   Insertion Magnetic Flow Meter Applications Guidelines

Recommended	Not Recommended
Raw water	Fluids where coating of probe can occur
Secondary effluent	Scum
Finished water	Raw sewage
	Sludges

Principle of Operation

Insertion magnetic flow meters operate by using Faraday's principle of electromagnetic induction in which the induced voltage generated by an electrical conductor moving through a magnetic field is proportional to the conductor's velocity. Figure 2.17 shows an insertion meter probe.

The induced voltage is received by the two electrodes mounted in the probe face. This signal is sent to the converter/transmitter where it is summed, referenced, and converted from a magnetically induced voltage to the appropriate scaled output. The unit uses pulsed-dc coil excitation.

Unlike the magnetic flow meter which measures the average fluid velocity,

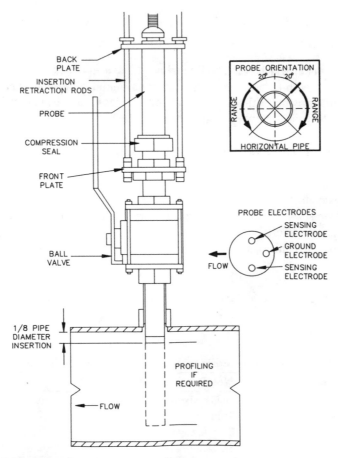

Figure 2.17.   Insertion magnetic flow meter.

the insertion magnetic flow meter measures the fluid velocity at the probe head.

Since the insertion meter measures the velocity at the probe, it is important to place the probe head at a point in the pipe where the flow is at mean velocity.

Figure 2.18 shows the velocity profile in a pipe for various Reynolds numbers. The Reynolds number is determined by:

$$R_n = V_m D/v$$

Where: R_n = Reynolds number
V_m = Mean velocity
D = Pipe diameter
v = Kinematic viscosity of fluid

R_n less than 3000 represents laminar flow. When R_n is above 4000, the flow is turbulent.

Nikuradse's equation for velocity distribution in a pipe is given by:

$$V_y/V_c = (Y/R)^{1/n}$$

Where: V_y = Velocity at position Y in pipe
V_c = Velocity at center of pipe
Y = Distance from pipe well
R = pipe radius
n = exponent dependent on Reynolds number

The exponent n has been determined empirically and is given for a range of Reynolds numbers (1). The variation in mean velocity over a wide flow range is demonstrated by the following examples:

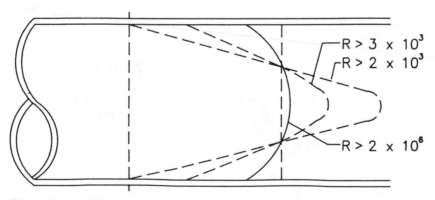

Figure 2.18.   Velocity profile vs Reynolds number.

$$R_n = 5000 \qquad n = 5 \qquad Y/R = 0.249$$
$$R_n = 3 \times 10^6 \qquad n = 10 \qquad Y/R = 0.237$$

The preceding demonstrates that if the flow is turbulent and fully developed, and if the tip of the probe is placed at 1/8 pipe inside diameter, then the mean velocity will stay within the magnetic field over a 10:1 flow range.

If a normal flow profile cannot be expected due to an upstream obstruction, a flow profile under actual operating conditions should be developed. The mean velocity point can be determined from a plot of insertion position versus flow reading.

You must use an area weighted mean velocity. Measure the velocity in a set of concentric rings. The sum of the velocity times ring areas divided by the total pipe area is the weighted mean velocity. In tests, the weighted mean average was 0.7 m/s (2.2 ft/s) while the mean average velocity was 1 m/s (3.2 ft/s).

Accuracy and Repeatability

Accuracy	$\pm 2\%$ of full scale if positioned at the mean velocity point
Repeatability	$\pm 1\%$ of full scale, 8 inch line size or less
	$\pm 2\%$ of full scale, up to 36 inch line size

At velocities below 1 m/s (3 ft/s) the accuracy falls off quickly because the flow becomes laminar and the velocity profile changes. Errors of as much as $\pm 4\%$ can be expected.

Flow-disturbing piping obstructions located too near the meter inlet and outlet may add an additional 1–10% of uncertainty to the measured flow. Avoid locating the following obstructions nearer than five pipe diameters to the meter inlet or outlet:

1. valves, gates
2. tees, elbows
3. pumps
4. severe reducers and expanders (greater than 30° included angle)

Meter orientation leading to a nonfull meter pipe (i.e., trapped gases) or resulting in material buildup on the electrodes severely degrades accuracy.

Manufacturers' Options

1. Probe
   a. stainless steel with epoxy face
   b. CPVC

      c. KYNAR
  2. Electrode materials
      a. 316 stainless steel
      b. titanium
      c. tantalum
      d. Hastelloy
  3. Transmitter
      a. auto zero calibration
      b. output signal: 4–20 mAdc, digital pulse, scaled digital pulse
      c. face-mounted indicating meter

Installation

- Locate the meter on the discharge side of pumps and on the upstream side of throttling valves.
- Locate the meter in a straight run of pipe free of valves or fittings with a minimum of 10 diameters upstream to elbows and fitting, 25 diameters upstream to control valves, and 5 diameters downstream to obstructions.
- The process conduit must flow full of liquid.
- Pipe sizing is critical. Size the pipe to provide a fluid velocity of 1–9 m/s (3–30 ft/s).
- Properly ground all insertion meters. Always ensure that the plant electrical system ground near the meter location provides adequate grounding. If a plant-wide grounding grid is available, ground the meter to it.
- Avoid locating meters near heavy induction equipment because it causes meter operational problems (100 HP motors and larger, no closer than 20 ft).
- Mount the probe so that the ground electrode is upstream.
- Provide sufficient space to facilitate calibration, in-line maintenance, or meter removal.
- Orient the meter so the electrodes lie in a plane perpendicular to the floor. See Figure 2.7.
- Use driven-shield signal leads and route them between the transmitter and meter through dedicated 2 cm (3/4 in.) conduit. Route power wiring in separate conduit.
- Wire power to the transmitter/converter and the coil driver through the same dedicated circuit. If separate circuits are used for transmitter/converter power and coil driver power, both circuits must originate from the same phase of the primary power feed.
- If the meter is mounted in a vertical pipe run the flow direction should be upward.

- Metered lines should not self-drain when shut down.

Designer Checklist

If you can answer "yes" to the following questions when designing or reviewing mag meter applications, the application should be correct.

- Does the process liquid to be measured have a measured conductivity greater than 5 μmhos per centimeter?
- Will the pipe flow full under ALL conditions, excluding shutdown?
- If intermittent flow is expected, will the meter remain full at no flow? Does the transmitter have low flow zero cut-out circuitry?
- Does the meter size ensure a flow velocity between 1.5–7.5 m/s (5–25 ft/s) for solids-bearing liquids or 1–9 m/s (3–30 ft/s), for nonsolids-bearing liquids?
- Are all piping elements/obstructions located at a minimum pipe diameters upstream or downstream of the meter?
- Is the meter grounded to a true ground?
- Is signal wiring between the transmitter and meter as specified by the meter manufacturer and routed in separate conduit?
- Has a dedicated power source been provided for the mag meter?
- Has the proper electrode material been selected to avoid excessive wear?
- Does the design provide environmental temperatures within the range specified by the manufacturer for both the meter tube and the transmitter/converter?

Maintenance and Calibration

Task	Frequency
1. Calibrate transmitter	Monthly
2. Flow calibrate the meter	Every three months

Deficiencies

The following problems could be encountered with insertion magnetic meter installations.

- Velocity skewing created by piping obstructions located too near to the meter causes accuracy problems.
- Pipe sizing does not maintain adequate flow velocities. Often this results from over-design for "future" flow rates.
- Improper installation results in nonfull pipe during low flows.

- Solids coating the electrodes due to low flow velocity or intermittent flow.
- Meter or transmitter located so that calibration and maintenance accessibility are difficult.
- Infrequent calibration.
- No provisions are made for meter calibration.
- Improper grounding.

BIBLIOGRAPHY

Magnetic

1. Liptak, B.G., and K. Venczel. *Instrument Engineers Handbook of Process Measurement*, rev. ed. (Radnor, PA: Chilton Book Company, 1982).
2. Kulin, G. "Recommended Practice For The Use Of Electromagnetic Flow Meters In Wastewater Treatment Plants." EPA 600/2-84-187 (Cincinnati, OH: U.S. Environmental Protection Agency, November, 1984).
3. Instruction Bulletin No. 10D1435A, rev. 1 (Warminster, PA: Fisher & Porter Company, 1969).
4. "Magnetic Flow Meter—Basic Theory," Product Data, PDS-15E001 Issue 3 (Rochester, NY: Sybron/Taylor Corp.).
5. "Magnetic Flow Meter—Application," Product Data, PDS-15E002 Issue 3 (Rochester, NY: Sybron/Taylor Corp.).
6. "Magnetic Flow Meter—Installation," Product Data, PDS-15E003 Issue 2 (Rochester, NY: Sybron/Taylor Corp.).

Sonic

1. Liptak, B. G., and K. Venczel. *Instrument Engineers Handbook of Process Measurement*, rev. ed. (Radnor, PA: Chilton Book Company, 1982).
2. Brown, A. E. " Application of Flowmeters to Water Management Systems," paper presented to Instrument Society of America, ISA/81 Conference, Anaheim, CA, October, 1981.
3. Powell, D. J. "Ultrasonic Flowmeters, Basic Design, Operation and Criteria Application," *Plant Engineering* (May, 1979).
4. Hall, J. "Choosing a Flow Monitoring Device," *Instruments and Control Systems* (June, 1981).

Turbine

1. Liptak, B. G., and K. Venczel. *Instrument Engineers Handbook of Process Measurement*, rev. ed. (Radnor, PA: Chilton Book Company , 1982).
2. Technical Bulletin No. TI, 16–6a. (Foxboro, MA: Foxboro Company, January, 1971).

Venturi Tubes and Flow Tubes

1. Liptak, B. G., and K. Venzcel. *Instrument Engineers Handbook of Process Measurement*, rev. ed. (Radnor, PA: Chilton Book Company, 1982).
2. Spink, K. L. *Principles and Practice of Flow Meter Engineering*, 9th ed. (Foxboro, MA: The Foxboro Company, 1967).
3. "Instrumentation in Wastewater Treatment Plants," WPCF Manual of Practice No. 21, (Washington,DC: Water Pollution Control Federation, 1978).
4. "Measurement of Fluid Flow by Means of Orifice Plates, Nozzles and Venturi Tubes Inserted in Circular Cross-Section Conduits Running Full," ISO/DIS 5167, draft revision of R781. (New York:International Standards Organization, 1976).
5. "Fluid Flow in Closed Conduits–Connections for Pressure Signal Transmissions Between Primary and Secondary Elements," ISO 2186. (New York: International Standards Organization, 1973).
6. "Standard Method of Flow Measurement of Water by the Venturi Meter Tube," ASTM D2458–69. (Philadelphia: American Society for Testing and Materials).
7. Henson, J. E. *Process Instrumentation Manifolds*. Research Triangle Park, NC: Instrument Society of America, 1981.
8. "Fluid Meters, Their Theory and Application," 6th ed., Report ASME Research Committee on Fluid Meters (New York: American Society of Mechanical Engineers, 1971).
9. Sprenkle, R. E. Piping Arrangements for Acceptable Flowmeter Accuracy. *ASME Transactions* 67:345 (1945).
10. Starret, P. S., P. F. Halfpenny and H. B. Noltage. "Survey of Information Concerning the Effects of Nonstandard Approach Conditions Upon Orifice and Venturi Meters," paper presented at Annual Winter Meeting, American Society of Mechanical Engineers, New York, NY, 1965.

Vortex Shedding

1. Ginesi, D., and Grebe, G. "Flow, A Performance Review," *Chem. Eng.*, (June 22, 1987), p. 108.
2. Swingwirl II Vortex Sales Bulletin, 6–87, Greenwood, IN: Endress + Hauser.

Insertion Magnetic

1. Schlechting, H., *Boundary Layer Theory*, 7th ed. (New York: McGraw-Hill Book Company, 1979).
2. Magnetic Flow Meter—Installation and Operation Manual," rev. D (Naperville, IL: Dynasonics, Inc.).
3. Moussiaux, J. J.,"Flow Determination and Field Calibration of Pulp Stock Process Flow Using a Point of Measurement Electromagnatic Flow Meter," in *Proceedings of the 1982 Joint Symposium* (Columbus, OH: April, 1982).

Chapter 3

Flow Measurement, Closed Conduit Gas Flow

As in the case of liquid flow meters, there are many different types of gas flow meters. Technologies include mechanical, differential pressure, vortex shedding, and mass rate of cooling.

This chapter includes orifice plates, venturi tubes, averaging pitot tubes, turbines, and mass flow meters. Vortex shedding, V-element, target, and positive displacement meters can also be used to measure gas flow. They are not discussed in this chapter however. See Chapter 2 for a discussion of V-element and vortex shedding meters.

ORIFICE PLATE

Application

Orifice plate gas flow meters are suitable for the application under the following general conditions:

1. The fluid is clean gas or steam.
2. A relatively large head loss is acceptable.
3. The Reynolds number at minimum flow is greater than 30,000.
4. A meter maximum to minimum ratio of 3:1 is acceptable.

Table 3.1 contains application guidelines for orifice plate gas flow meters.

The discussions in this section pertain to concentric orifice plates as shown in Figure 3.1. Other configurations, segmental and eccentric, are available to accommodate particular application problems.

Principle of Operation

Orifice plates are differential-producing head-type flow measuring devices made of a thin, flat plate having an opening (orifice). When installed in the cross section of the process pipe, orifice plates cause an increase in the flow velocity as the process gas moves through the orifice, causing a corresponding decrease in downstream pressure. A differential pressure measuring device is connected across the orifice plate to sense the differential pressure. Figure 3.2 illustrates this principle.

Proper sizing of the orifice plate is required for accurate flow measurement. Use the following guidelines:

Plate Thickness

1. Pipe I.D. from 5–20 cm (2–12 in.): 0.3 cm (1/8 in.) thick. Pipe I.D. 35 cm (14 in.) and larger: 0.6 cm (1/4 in.) thick.
2. Bevel plates thicker than 0.3 cm (1/8 in.) on the downstream orifice edge as shown in Figure 3.1.

Orifice Diameter

The Beta ratio is defined as the ratio of orifice diameter (d) to the pipe I.D. (D) and is critical for accurate flow measurement. For air and steam flow measurement, it is recommended that the Beta ratio be greater than 0.2 but less than 0.7. The flow calculations used for determining the Beta ratios are standardized but rather complex. These calculation methods are thoroughly covered in Bibliography numbers 2, 5, and 7.

The orifice plate flow meter is a primary sensing element which creates a pressure differential proportional to the square of the flow rate. This pressure

Table 3.1. Orifice Plate Application Guidelines

Recommended	Not Recommended
Boiler steam	Wet steam
Compressed digester gas	Low pressure (uncompressed)
Natural gas	digester gas
Air	Strongly corrosive gases
Oxygen	
Draft air	

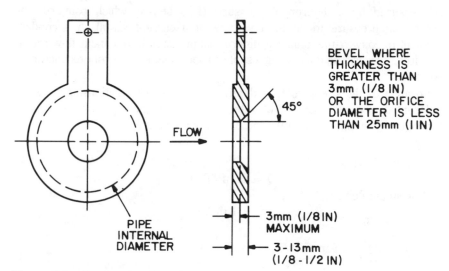

Figure 3.1. Concentric orifice plate.

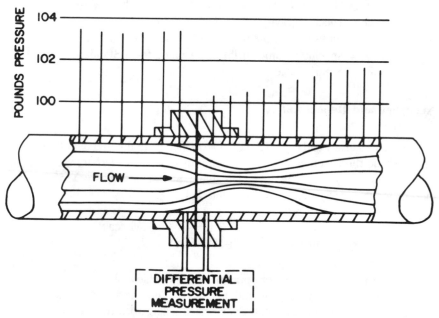

Figure 3.2. Pressure profile.

is measured by a differential pressure (DP) sensor which converts the differential pressure to either a voltage or a current signal. A secondary element to convert the nonlinear differential pressure into a linear flow rate is required. This can be done using a square root scale, square root extractor, or computer.

Location of the pressure taps determines the exact relationship between differential pressure and flow rate. These relationships are stated in Bibliography numbers 1 and 9. The basic equations are:

For gas flow:
$$Q = K(hP/TG)^{1/2}$$

For steam flow:
$$W = K(h/V)^{1/2}$$

Where Q = volume flow rate (scfh)
  K = basic orifice expansion, flow and conversion factors that usually are constant for a given application
  h = differential pressure across the orifice in inches of water
  P = absolute flowing pressure, psia
  T = absolute flowing temperature (°R = °F + 460)
  G = specific gravity of gas (air = 1.0)
  W = mass flow in pounds per hour (lb /hr)
  V = specific volume (ft/lb) determined from Standard Steam Tables

Accuracy and Repeatability

The total accuracy and repeatability of an orifice plate flow-measuring system must include the accuracy and repeatability of the orifice, DP sensor, and the square root extractor. The following limits are achievable by orifice plate meters for the applications listed in Table 3.1:

Accuracy: ± 2 to $\pm 5\%$ of full scale
Repeatability: $\pm 1\%$ of full scale

These accuracy levels reflect optimum values achievable by the measuring system when properly applied and installed. The main factors which will degrade these levels include: improper orifice sizing with respect to the Beta ratio; flows either less or greater than anticipated; and piping configurations which disrupt the velocity profile.

Manufacturers' Options

1. Type of construction materials
2. Pressure connection location
   a. flange taps (standard)
   b. vena contracta taps
   c. radius taps
   d. corner taps
3. Orifice shape and location:
   a. concentric (standard), with or without drain and vent holes
   b. eccentric, with or without drain and vent holes
   c. segmental, with or without drain and vent holes
4. Removable orifice plate (without process disruption)
5. Secondary element:
   a. differential pressure transmitter (standard)
   b. manometer transmitter (not recommended)
6. Additional sensors (for measuring line temperature and pressure) and a module for calculating gas flow in standard units

Installation

Primary System

- Mount the orifice plate in either horizontal or vertical process piping. Pressure tap locations and DP transmitter locations will differ according to the orientation selected.
- Install beveled or cutaway plates (Figure 3.1) with the flat surface upstream.
- Use 1.6 millimeter (1/16 in.) thick gaskets, graphited on the side next to the plate. The gaskets must not extend into the pipe or obstruct vent and drain holes (if used).
- Provide straight run smooth piping upstream and downstream of the orifice plate. The length of straight run required depends on the Beta ratio. Recommended lengths are shown in Figure 3.3. Use straightening vanes when it is not practical to install the meter with the recommended straight pipe length.
- Pressure taps should be free of any burrs or protrusions into the pipe.

Secondary System

Installation of the secondary system (pressure connections, tap line run, and DP transmitter location) differs between applications for steam flow (condensable) and gas flow measurement.

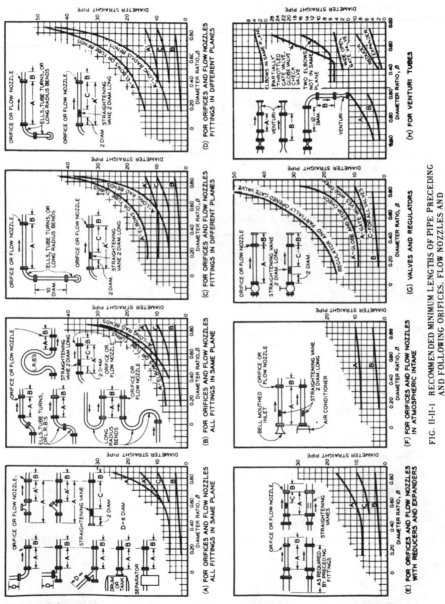

FIG. II-II-1   RECOMMENDED MINIMUM LENGTHS OF PIPE PRECEDING AND FOLLOWING ORIFICES, FLOW NOZZLES AND VENTURI TUBES (ALL CONTROL VALVES, INCLUDING REGULATORS, SHOULD BE LOCATED ON OUTLET SIDE OF PRIMARY ELEMENT.)

Figure 3.3.   Orifice straight run requirements (reprinted courtesy of ASME).

- Steam flow measurement.
  - a. A typical installation diagram is shown in Figure 3.4.
  - b. Always use condensing chambers on tap line runs. Mount each chamber at the same level.
  - c. Install horizontal portions of tap line runs so they slope downward from the orifice at a 1:12 slope.
  - d. Install the DP transmitter below the orifice plate location for both vertical and horizontal piping runs.
  - e. If the transmitter must be mounted above the pressure connections either vena contracta or pipe taps are recommended. Flange taps are not recommended.

- Gas flow measurement.
  - a. A typical installation diagram is shown in Figure 3.5.
  - b. Mount the DP transmitter above the orifice plate for both vertical and horizontal piping runs.
  - c. If the gas is corrosive, use a liquid seal with diaphragm pressure connections to isolate the transmitter.

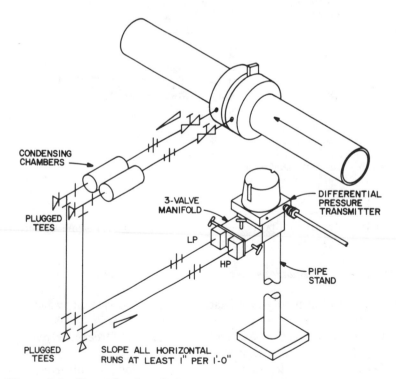

Figure 3.4. Steam flow installation.

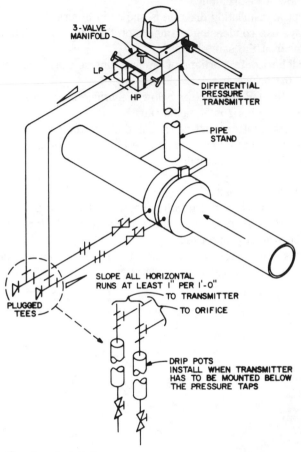

Figure 3.5. Gas flow installation.

d. Install horizontal portions of tap line runs so they slope upward from the orifice at a 1:12 slope.

Additional Recommendations

- Locate the transmitter to facilitate easy access for calibration and maintenance.
- Tap line runs should not exceed 15 m (50 ft) and, if freezing is possible, should be insulated or heat traced.
- Take special care to mount the DP transmitter pressure connection plumbing so the differential measurement is not affected.

- An isolation valve manifold and quick disconnects should be installed in the tap lines to facilitate sensor calibration.

Designer Checklist

If you can answer "yes" to the following questions when designing or reviewing orifice plate gas flow meter applications, the application should be correct.

General Items

- Is the process gas or steam recommended in Table 3.1?
- Is a large head loss acceptable?
- Is the Reynolds number at minimum flow greater than 30,000?
- Has the proper orifice size based on Beta ratio, (d/D ratio) been determined for the expected flow range and allowable pressure loss?
- Is the Beta ratio greater than 0.2 and less than 0.7?
- Will the meter construction materials withstand the corrosive properties of the fluid to be measured?
- Has the proper differential range been selected for the DP transmitter?
- Have the flange gaskets been properly sized to insure no protrusion into the inside diameter of the process pipe?
- Does the straight run piping conform to the minimum requirements in Figure 3.4?
- Are the tap line runs less than 15 m (50 ft) long?
- If freezing is possible, are the tap lines adequately insulated or heat traced.
- Have an isolation valve manifold and quick disconnects been installed in the tap lines to facilitate DP sensor calibration?
- Will the DP sensor be mounted in a vibration free location?

Steam Flow Measurement

- Have condensing chambers been provided in the tap lines and are they of adequate size?
- Is the transmitter mounted below the process connections? If it is not, have either vena contracta or pipe taps been used rather than flange taps? Insulate from thermal shocks with tube loop (pig tail) or other means.

Gas Flow Measurement

- Is the transmitter mounted above the process connections?
- Have condensate traps been installed at the lowest point of the tap line runs?
- If the gas is corrosive, have diaphragm sealed pressure connections been provided?

Maintenance and Calibration

Primary System (Orifice Assembly)

Task	Frequency
1. Test the primary with a portable manometer	Monthly

If accuracy problems persist, remove orifice plate and inspect orifice for solids buildup and/or wear.

Secondary System (DP Sensor)

Task	Frequency
1. For gas flows, empty condensate traps	Weekly
2. Recalibrate trasmitter using a portable manometer or other suitable calibration test set	Monthly

Deficiencies

The following problems are commonly encountered with orifice plate gas flow meter applications.

- Orifice oversized (Beta ratio too high), thus generating a differential too low to be accurately monitored by the DP transmitter provided.
- Orifice properly sized, but the wrong DP sensor range was selected.
- Condensate traps not provided, causing water accumulation and occasional freezing in the lines.
- Unequal tap line lengths or elevations, causing differential measurement errors.
- Insufficient straight run piping provided.
- Insufficient calibration of primary and secondary systems.

VENTURI TUBES AND FLOW TUBES

Applications

Venturi and proprietary flow tube gas flow meters are suitable for application under the following general conditions:

1. The Reynolds number of the process stream at the meter is greater than 100,000. 150,000 is preferable.
2. A meter rangeability of 3:1 is acceptable.

Application guidelines for venturi tubes and flow tubes are shown in Table 3.2.

Principle of Operation

Venturi Tube

A venturi tube operates on the principle that a gas flowing through a meter section containing a convergence and constriction of known shape and area will cause a pressure drop at the constriction area. The difference in pressure between the inlet and the constriction area (throat) is proportional to the square of the flow rate. Figure 3.6 shows a cutaway of a typical venturi tube.

Flow Tube

Several manufacturers provide differential-causing flow tubes that are modified versions of the classical venturi tube. These devices operate on the same principle as the classical venturi. However, they provide features which make them more attractive for some applications, i.e., less space is required for installation and overall head loss is reduced. Figure 3.7 shows three commonly used flow tubes.

Both the venturi tube and the proprietary flow tubes are primary elements and require a secondary element to sense the pressure differential and convert

Table 3.2.   Venturi Tubes and Flow Tubes Application Guidelines

Recommended	Not Recommended
Boiler stream	Any low-pressure (uncompressed)
Compressed digester gas	gas flows
Incinerator draft/blower air	
Air flow	
Oxygen flow	
Corrosive gases	

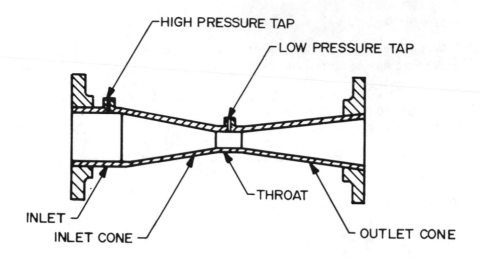

Figure 3.6.   Classic venturi tube.

it to a usable signal. The secondary element in most applications is a differential pressure (DP) transmitter.

The DP produced by a flow tube is representative of the volumetric flow rate at the actual operating temperature and pressure. Appropriate temperature and pressure sensors are required to correct the transmitter output to standard reference conditions.

The DP in the tube is measured at the inlet and throat (Figure 3.6). One DP measurement method uses single connections in the inlet and throat of the tube. The pressure tap lines are coupled directly to the tap holes and run to the DP transmitter.

As an alternative method, piezometric rings may be used to sense inlet and throat pressures. These consist of several holes around the circumference of the tube at the inlet and throat tap locations. Each set of holes is connected to an annular ring to give an average of the pressure at each tap hole connected to the ring. Piezometric rings are usually used in large diameter tubes to minimize velocity profile skewing.

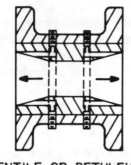

GENTILE OR BETHLEHEM
FLOW TUBE

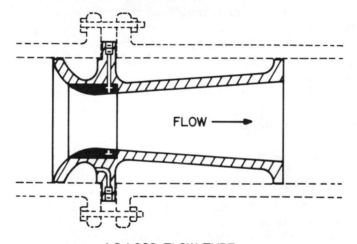

FLOW ——➤

LO-LOSS FLOW TUBE

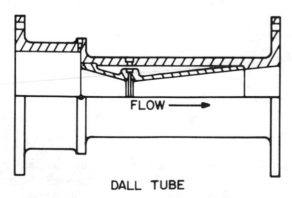

FLOW ——➤

DALL TUBE

Figure 3.7. Proprietary flow tubes.

Accuracy and Repeatability

The accuracy and repeatability of these meters vary with the type used. The characteristics of the secondary element (transmitter) must also be included in the total accuracy figure. The following limits can be expected when considering these types of flow meters for applications previously listed:

Accuracy:	$\pm 1\%$ of actual flow
Repeatability:	$\pm 1\%$ of acutal flow

The values shown are for a complete system (primary and secondary) when properly applied and installed. Factors which will degrade these levels during operation include: improper tube sizing, improper DP range, insufficient straight pipe before and after the meter and piping elements which disrupt the velocity profile.

Venturi tubes, flow tubes, and flow nozzles are all capable of exceeding the accuracy values shown. Consult the manufacturer if greater accuracy is required. Caution: both the primary and secondary element must be considered when designing for optimum accuracy.

Manufacturers' Options

Primary Meter Tube

1. Single sensing ports
2. Piezometer ring sensing
3. Inspection openings
4. Manual rodders for cleaning sensing ports

Secondary Element (Transmitter)

1. Differential pressure transmitter (see Chapter 6)
2. Manometer transmitter (not recommended)
3. Temperature and pressure correction system

Installation

Primary System

- Venturi and flow tubes may be installed in any position to suit the requirements application. However, the primary and secondary system must be accessible for maintenance and calibration.
- Flow disturbing obstructions, pipe fittings and valves will produce

meter inaccuracies and should not be located too near the meter inlet. Use Figure 3.8 to determine the minimum upstream distance of straight pipe recommended between fittings and the meter inlet.
- If a flow control valve is required in the line, it should be placed downstream from the meter tube as shown in Figure 3.8.
- Locate all downstream pipe fittings a minimum distance of two pipe diameters downstream of the throat tap(s).
- Install the primary and secondary flow elements in an accessible location with suitable space provided for maintenance and calibration.

Secondary System

Installation of the secondary system (pressure connections, tap line run, and DP transmitter location) differs between applications for steam flow (condensable) and gas flow measurement.

For steam flow measurement the tap lines are normally flooded with condensate.

- Use condensing chambers on tap line runs. Mount each chamber at the same level.
- Slope horizontal tap line runs downward from the pressure taps at a 1:12 slope.

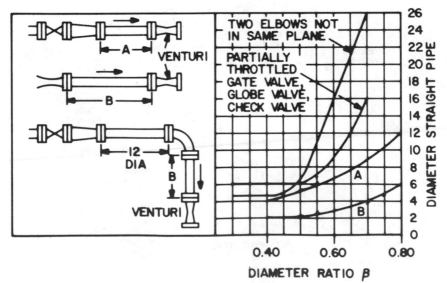

Figure 3.8.   Venturi piping requirements.

- Install the DP transmitter below the process pipe for horizontal runs, or below the pressure connections for vertical runs.

A typical installation diagram is shown in Figure 3.9. For gas flow measurement the tap lines are normally dry.

- Mount the DP transmitter above the process pipe for horizontal runs, or above the pressure connections for vertical runs.
- If the gas is corrosive, use a liquid seal with diaphragm pressure connections to isolate the transmitter.
- Slope horizontal tap line runs upward from the pressure taps at a 1:12 slope.

A typical installation diagram is shown in Figure 3.10.

Additional Recommendations

- Locate the transmitter to provide easy access for calibration and maintenance.

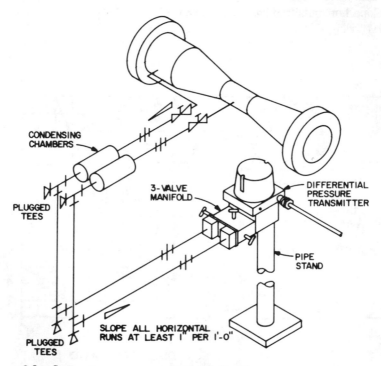

Figure 3.9. Steam flow measurement installation diagram.

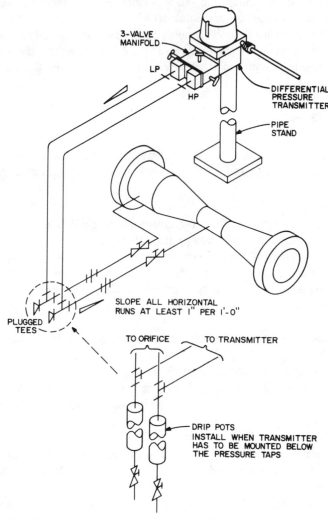

Figure 3.10. Gas flow measurement installation diagram.

- Tap line runs should not exceed 15 m (50 ft.) in length.
- If a freezing potential exists, heat trace and insulate the tap lines.
- Mount the transmitter so that the high and low pressure connections are at exactly equal elevation. Failure to do so will create a bias in the indicated DP which will in turn introduce an error in flow measurement.
- An indicator gauge (DP) should be placed near the primary element for

convenience in calibration and performance monitoring (usually mounted on the secondary element).

- The Beta ratio and overall diameter of the flow tube should be carefully determined for the expected flow range. Accordingly, the range of the DP transmitter must match that of the flow tube over the expected flow range.

Designer Checklist

If you can answer "yes" to the following questions when designing or reviewing venturi tube gas flow meter applications, the application should be correct.

- Is the process gas or steam recommended in Table 3.2?
- Is the Reynolds number greater than 150,000?
- Has the primary element been properly sized to generate a suitable differential pressure over the range of the expected flow? Meter sizing for 20-year projected flow typically results in oversizing.
- Has the proper differential range been selected for the secondary device?
- Has adequate straight-run piping been provided both up and down-stream?
- Are the tap lines sloped properly?
- Is steam flow being measured? If so, is the transmitter installed below the process connections and have condensing chambers been provided in the tap lines?
- Are the tap line runs less than 15 m (50 ft)?
- Is freezing a possibility? If so, are the tap lines heated and insulated?
- Are both the primary and secondary systems readily accessible for maintenance?
- When metering gas flow, is the transmitter installed above the process connections?
- When metering gas flow, have condensate traps or drip legs been installed at the low point of the tap line runs?

Maintenance and Calibration

Primary System

If performance monitoring indicates a large accuracy loss, test the primary with a portable manometer.

Secondary System

Task	Frequency
1. Bleed off condensate in the tap lines	Weekly at first until history is established
2. Zero and recalibrate transmitter using a portable manometer	Monthly

Deficiencies

The following problems are commonly encountered with existing venturi and flow tube applications:

- Meter oversized. Low flow differentials are lost due to square root function cutoff.
- Gas buildup caused by tap lines that are not properly sloped or not equipped with bleed valves.
- Improper differential range selection for the DP transmitter.
- Tap lines are plugged or restricted.

AVERAGING PITOT TUBES

Applications

Averaging pitot tubes are suitable for the application under the following general conditions:

1. The process is clean gas or steam (free of solids).
2. You want to minimize head loss.
3. A meter rangeability of 3:1 is acceptable.
4. The Reynolds number of the fluid is greater than 40,000.

Table 3.3 shows application guidelines for averaging pitot tubes.

Principle of Operation

Averaging pitot tubes are differential-producing flow measuring devices that consist of an insertion probe with multiple upstream sensing ports and a single downstream static port. The probe is geometrically constructed such that an average upstream pressure is measured. Figure 3.11 shows the probe.

The averaging pitot tube, like the venturi meter is a primary sensing element that creates a pressure differential proportional to the square of the flow rate.

Table 3.3. Averaging Pitot Tubes Application Guidelines

Recommended	Not Recommended
Boiler stream	Gas or steam with particulate solids
Compressed digester gas	Low pressure (uncompressed digester gas)
Natural Gas	Corrosive gases
Blower air	
Oxygen	
Incinerator draft air	

Like the venturi, a DP cell is used to measure the differential pressure and convert it to a voltage or current signal. A secondary element to convert the nonlinear differential into a linear flow rate is required. In most cases, this device is a square root extractor.

Accuracy and Repeatability

The accuracy and repeatability of the averaging pitot tube is good. However, the characteristics of the DP transmitter and square root extractor must be included in the total accuracy figure. The following performance limits are attainable by averaging pitot tubes:

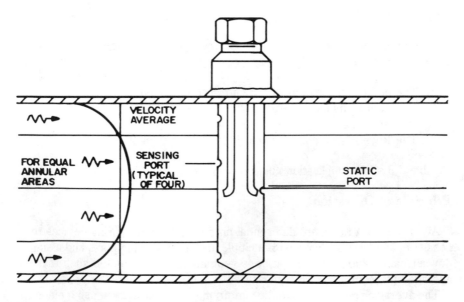

Figure 3.11. Pitot tube probe.

Accuracy: ±2 to ±5% of full scale
Repeatability: ±1% of full scale

These accuracy levels reflect optimum values achievable for proper application and installation. Factors that will degrade these levels include operation at actual flows outside the expected flow range, and piping elements that disrupt the velocity profile.

Manufacturers' Options

1. Mounting:
   a. Mounting coupling
   b. Flange
   c. Hot tap
2. Secondary element:
   a. Differential pressure transmitter (standard)
   b. Manometer transmitter (not recommended)
3. Calculation modules for standard pressure, temperature correction and linearization to flow units.

Installation

Primary System

- Mount the averaging pitot tube in either horizontal or vertical process piping. However, the tap line configuration and DP transmitter locations will differ.
- Install the probe with the multiple ports facing upstream.
- Provide adequate straight-run smooth piping upstream and down-stream of the pitot tube. When it is not practical to install the pitot tube with the recommended straight pipe length, use straightening vanes. Recommended lengths with and without vanes are shown in Figure 3.12.

Secondary Systems

Installation of the secondary system (pressure connections, tap line run, and DP transmitter location) differs between applications for steam flow (condensable) and gas flow measurement.

A typical installation diagram with correct orientation for steam flow measurement shown in Figure 3.13.

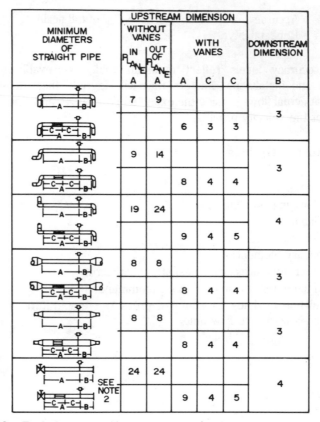

Figure 3.12.   Typical upstream/downstream requirements.

- Always use condensing chambers on tap line runs. Mount each chamber at the same level. Size condensation chambers large enough for the application to prevent flooding between routine maintenance checks.
- Install horizontal portions of tap line runs so they slope upward from the primary element at a 1:12 slope.
- Install the DP transmitter below the process pipe for horizontal piping runs; below the pitot tube location for vertical piping runs.

A typical installation diagram with correct orientation for gas flow measurement is shown in Figure 3.14.

- Mount the DP transmitter above the process pipe for horizontal piping runs and above the pitot tube for vertical piping runs.

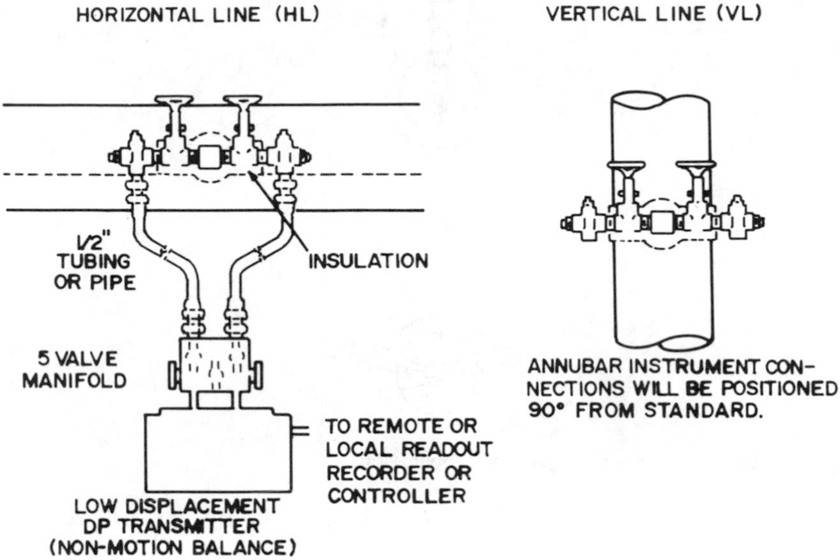

HORIZONTAL LINE (HL)   VERTICAL LINE (VL)

1/2"
TUBING   INSULATION
OR PIPE

5 VALVE   ANNUBAR INSTRUMENT CON-
MANIFOLD   NECTIONS WILL BE POSITIONED
                                             90° FROM STANDARD.

                          TO REMOTE OR
                          LOCAL READOUT
                          RECORDER OR
LOW DISPLACEMENT   CONTROLLER
DP TRANSMITTER
(NON-MOTION BALANCE)

Figure 3.13.   Orientation for steam flow applications.

- Install horizontal portions of tap line runs so they slope upward at a minimum of a 1:12 slope.

Additional Recommendations

- Locate the transmitter to permit easy access for calibration and maintenance.
- Tap line runs should not exceed 15 m (50 ft). If freezing is possible, insulate and heat trace tap lines.
- Mount the transmitter so that the high and low pressure connections are at exactly equal elevation.
- Provide a 3-valve manifold and quick connects for connecting a manometer during calibration.

Designer Checklist

If you can answer "yes" to the following questions when designing or reviewing pitot tube gas flow meter applications, the application should be correct.

Figure 3.14.   Orientation for gas flow applications.

General Items

- Is the process gas or steam recommended in Table 3.3 and free of particulates?
- Has the proper differential pressure range been selected for the secondary device, i.e., DP transmitter?
- Has adequate straight-run piping been provided to conform to the minimum requirements in Figure 3.12?
- Do the horizontal tap line runs slope upward at a minimum of a 1:12 slope?
- Are the tap line runs less than 5 m (15 ft) long?
- If freezing is possible, are the tap lines insulated and heat traced?
- Have a 3-valve manifold and quick disconnects for a manometer been provided for calibration?
- Will the meter be mounted in a vibration-free location?

Steam Flow Measurement

- Have condensing chambers been provided in the tap lines and are they of adequate size?
- Is the transmitter mounted below the process connections?

Gas Flow Measurement

- Is the transmitter mounted above the process connections?
- Have condensate traps been installed at the lowest point of the tap line runs?

Maintenance and Calibration

Primary System

Task	Frequency
1. Test the primary with a portable manometer	When performance monitoring indicates large accuracy loss
2. Remove pitot tube and inspect orifice for solids buildup or wear	When performance monitoring indicates extended periods of poor accuracy

Secondary System

Task	Frequency
1. Empty condensate traps	Weekly
2. Check the transmitter calibration using a portable manometer or other suitable calibration test set	Monthly

Deficiencies

The following problems are commonly encountered with existing pitot tube gas flow meter applications.

- The wrong range on the DP transmitter was selected.
- Condensate traps were not provided, causing water accumulation and occasional freezing in the lines.
- Unequal tap line lengths or elevations used, causing differential errors.
- Insufficient straight-run piping provided.
- Infrequent maintenance.

TURBINE FLOW METERS

Application

Turbine meters are suitable to measure gas flow under the following general conditions::

1. An intermittent flow may be expected.
2. The process fluid is relatively clean.
3. A high head loss is acceptable.

Principle of Operation

Turbine flow meters consist of a pipe section with a multibladed rotor suspended in the fluid stream on a free running bearing, (see Figure 3.15). The plane of rotation of the rotor is perpendicular to the flow direction and the rotor blades sweep out nearly to the full bore of the pipe. The rotor is driven by the process gas impinging on the blades. Within the linear flow range of the meter, the angular velocity of the rotor is directly proportional to the liquid velocity which is, in turn, proportional to the volumetric flow rate. The speed of rotation is sensed by an electromagnetic pickup coil which produces a pulse. The output signal is a continuous voltage pulse train with each pulse representing a discrete volume of gas. The turbine output frequency is proportional to the volumetric flow rate at the actual operating temperature and pressure. An appropriate temperature and pressure correction system is required to convert the meter output into a volumetric rate at standard reference conditions. Turbine flow meter application guidelines are presented in Table 3.4.

Accuracy and Repeatability

When properly applied and installed, the accuracy and repeatability characteristics of turbine flow meters should be:

Accuracy: ± 0.5% of actual flow, within the linear
                 range of the meter

Table 3.4.   Turbine Flow Meter Application Guidelines

Recommended	Not Recommended
Stream	Low-pressure (uncompressed)
Compressed digester gas	digester gas
Natural Gas	
Air	

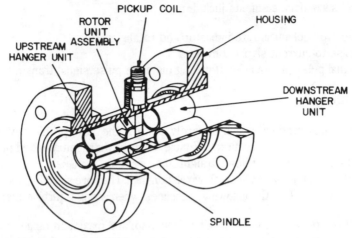

Figure 3.15. Gas turbine meter.

> Repeatability: ±0.05% within the linear range of the
> meter

Each turbine flow meter has a unique "K" factor (the number of pulses generated per unit volume) which is determined during factory flow calibration. This factor, and thus the accuracy of the meter, is affected by mechanical wear.

Manufacturers' Options

1. Wetted parts materials
   a. stainless steel (standard)
   b. Hastelloy C
   c. Teflon bearings
2. Flow straightening vanes
3. Additional electromagnetic pickup and associated electronics for increased accuracy.
4. Pressure and temperature correction system for calculating volumetric flow under standard conditions.

Turbine meters require secondary elements for indicating flow at the meter or retransmitting for remote monitoring. It is suggested these elements be purchased from the same manufacturer.

If another supplier is used, take care to ensure that both units are compatible with regard to signal pulse shape, amplitude, width, and frequency.

Typical secondary elements include:

1. Electromechanical rate indicator and totalizer
2. Pulse-to-current signal converter
3. Signal pulse preamplifier (for long distance pulse signal transmission)

Installation

- Piping obstructions which severely disturb the flow profile severely affect turbine meter accuracy. Figure 3.16 shows recommended piping installation, including flow straightening vanes.
- Turbine meters have a linear flow relationship. They are sized by volumetric flow rate. Use the following guidelines when sizing a turbine meter:

    a. Each meter size has a specified minimum and maximum range of flow linearity and should not be used for flow rates outside that range.
    b. The maximum flow rate for the application should be 70% to 90% of the maximum flow rate specified for the meter.
    c. Size the meter on actual volume flow and not on reference or standard units.
    d. The meter size should be less than the diameter of the process piping.

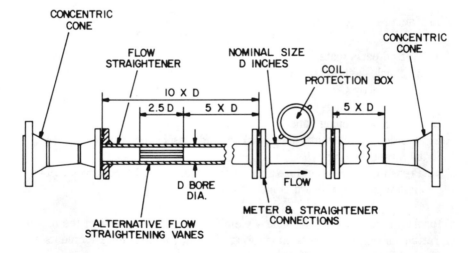

Figure 3.16. Meter installation.

  e. Available turbine meter sizes range from 0.5 – 60.0 cm (3/16 to 24 in.) in diameter.

- The recommended minimum upstream straight-run for optimum accuracy is 25 to 30 pipe diameters. If necessary, this distance may be reduced to 10 pipe diameters by installing straightening vanes. The following pipe fittings produce flow disturbances that will degrade meter accuracy if placed closer than the specified distances:
  a. valves and gates
  b. tees and elbows
  c. severe reducers and expanders (greater than 30-degree included angle)
- Locate downstream piping obstructions at least 5 pipe diameters from the meter.
- Use shielded cable between the turbine meter and secondary electronics.
- Route power wiring and signal cable in separate conduit.

Designer Checklist

If you can answer "yes" to the following questions when designing or reviewing turbine flow meter applications, the application should be correct.

- Is the intended process gas recommended in Table 3.4?
- Can the expected head loss be tolerated?
- Are all pipe obstructions located a minimum upstream distance of:
  a. 10 pipe diameters when flow straightening vanes are used?
  b. 25 to 30 pipe diameters without flow straightening vanes?
- Is there a minimum downstream distance of 5 pipe diameters to any flow disturbing fittings?
- Has the proper secondary flow indicator/signal conditioner been provided?
- Is power available for the secondary element?
- Is the meter accessible for maintenance?
- Is the meter diameter smaller than the process piping?
- If temperature and pressure correction are desired, have provisions been made?

Maintenance and Calibration

The meter constant "K" is determined at the manufacturer's facility prior to meter shipment and is performed under standard conditions. If the intended application process gas has significantly differing physical characteristics, the manufacturer should be consulted for additional testing data.

Turbine meters do not require calibration, but periodic calibration may be required on the secondary element. When meter accuracy becomes questionable (as observed through performance monitoring) check the "K" factor by physical testing.

Deficiencies

The following problems are commonly encountered with turbine meter applications:

- Inadequate upstream and downstream straight run piping, resulting in poor meter accuracy.
- Meter sized too large, resulting in poor accuracy at low flows.
- Meter and secondary element are not compatible because of differing electrical specifications.
- Although rare, the factory determined "K" factor is incorrect.
- Damage to bearings and nose cone from dirt or entrained liquids.

THERMAL MASS FLOW METERS

Applications

Thermal mass flow meters are suitable for the application under the following general conditions:

1. The process is gas or steam.
2. You want to minimize head loss.
3. The flow rate will vary widely 0.1 to 60 m/s (0.25 to 200 ft/s).
4. A very low velocity is expected.
5. There will be a wide range of operating temperatures.

Application guidelines for averaging thermal mass are in Table 3.5.

Principle of Operation

A thermal mass flow meter measures the mass flow rate of gas or steam. It consists of an insertion probe assembly composed of two pairs of thermowells. One of the thermowells in each pair houses a platinum Resistance Temperature Detector (RTD). The other half of one thermowell pair houses a heater which heats the adjacent RTD. This RTD is the active RTD. The second RTD acts as a reference sensor and its adjacent thermowell provides mass equalization for the heater. Figure 3.17 shows the probe. The configuration creates a

Table 3.5. Averaging Thermal Mass Application Guidelines

Recommended	Not Recommended
Natural gas	Gas or steam with water droplets
Blower air	Gas mixtures with widely variable
Oxygen	variable compositions
Boiler steam	
Incinerator draft air	
HVAC air	
Digester gas[a]	

[a]See Accuracy and Repeatability

temperature differential between the two RTDs which is greatest at zero flow. Changes in flow rate affects the heat dissipation and affects the temperature differential between the two RTDs. The temperature differential is a logarithmic function of the mass flow rate as shown in Figure 3.18. Because the flow-induced heat dissipation effect is a logarithmic function, a wide turndown with very good low flow sensitivity is available. A signal converter is necessary to convert the nonlinear signal into a linear mass flow rate.

Accuracy and Repeatability

The accuracy and repeatability of the mass flow meter is excellent over a wide range of flows. When properly applied and installed, the following performance limits are attainable:

Accuracy	$\pm 1\%$ of full scale at the calibration of 10:1 or less
	$\pm 3\%$ of full scale for larger turndowns
Repeatability:	$\pm 1\%$ of full scale

Factors that will degrade performance include operation at actual flows outside the expected flow range, piping elements that disrupt the velocity profile, and large changes in gas composition.

For digester gas, if the sensor is calibrated for a 70:30 mixture of methane:carbon dioxide and the mixture changes to 65:35, the reading will change by about 2%. If this accuracy is unacceptable, compensation circuitry can be provided. A carbon dioxide gas analyzer would be required.

Manufacturers' Options

1. Probe Assembly
   a. 316 stainless steel
   b. Hastelloy C

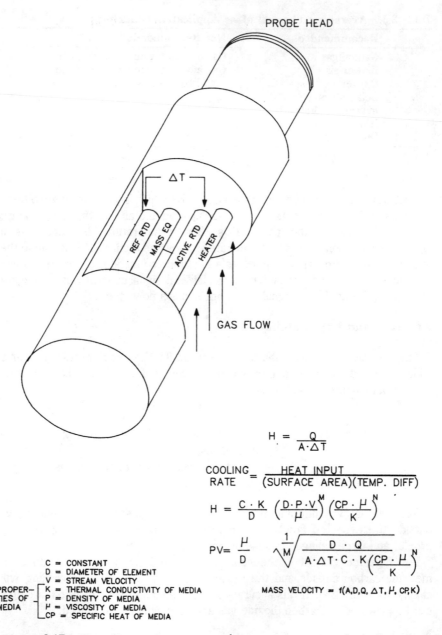

$$H = \frac{Q}{A \cdot \Delta T}$$

$$\frac{\text{COOLING}}{\text{RATE}} = \frac{\text{HEAT INPUT}}{(\text{SURFACE AREA})(\text{TEMP. DIFF})}$$

$$H = \frac{C \cdot K}{D} \left(\frac{D \cdot P \cdot V}{\mu}\right)^{M} \left(\frac{CP \cdot \mu}{K}\right)^{N}$$

$$PV = \frac{\mu}{D} \sqrt[\frac{1}{M}]{\frac{D \cdot Q}{A \cdot \Delta T \cdot C \cdot K \left(\frac{CP \cdot \mu}{K}\right)^{N}}}$$

C = CONSTANT
D = DIAMETER OF ELEMENT
V = STREAM VELOCITY
PROPER- ⌈ K = THERMAL CONDUCTIVITY OF MEDIA
TIES OF ⎹ P = DENSITY OF MEDIA
MEDIA ⎹ μ = VISCOSITY OF MEDIA
      ⌊ CP = SPECIFIC HEAT OF MEDIA

MASS VELOCITY = $f(A, D, Q, \Delta T, \mu, CP, K)$

Figure 3.17.   Thermal mass flow meter probe.

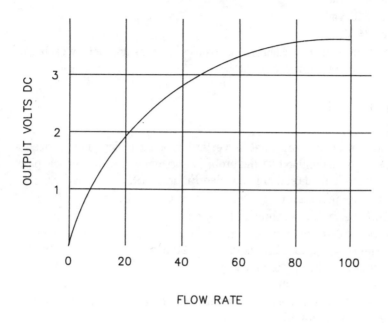

Figure 3.18. Logarithmic signal.

2. Transmitter Outputs
    a. 4–20 mAdc
    b. 10–50 mAdc
    c. 0–5 Vdc
    d. 0–10 Vdc
3. LCD display scaled in mass flow, volumetric flow or velocity
4. Temperature compensation if process temperature span is greater than 30°F.
5. Totalizer
6. Remote readout
7. Mounting
    a. 1 or 1–1/4" NPT
    b. flange
    c. hot tap
8. Housing
    a. NEMA 4
    b. NEMA 7

9. Power
    a. 115 Vac
    b. 220 Vac
    c. 24 Vdc
10. Calibrator unit to calibrate electronics. Probes are factory calibrated and cannot be field calibrated.

Installation

- Mount in either horizontal or vertical process piping. For horizontal pipes, side-mount so that the probe is horizontal. Use on vertical pipes only if the flow direction is up. See Figure 3.19.
- Install the probe in the center of the process line.
- Insert the probe to center of the pipe.
- Provide adequate straight-run smooth piping upstream and downstream of the probe. Provide at least 20 pipe diameters after the last flow disturbance upstream and 10 pipe diameters before the next disturbance downstream.
- Mount with flow across the sensing element in the direction which the probe was calibrated.

Designer Checklist

If you can answer "yes" to the following questions when designing or reviewing thermal mass flow meter applications, the application should be correct.

- Is the process gas or steam recommended in Table 3.5 and free of water droplets?
- Has adequate straight-run piping been provided? This meter is somewhat more sensitive than a turbine meter and less sensitive than a pitot tube.
- Will the meter be mounted in a vibration-free location?

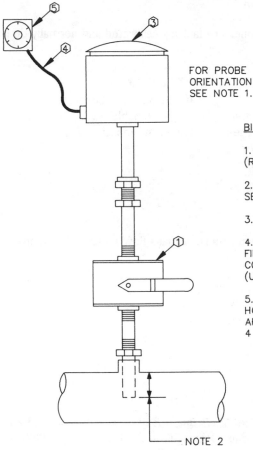

FOR PROBE
ORIENTATION
SEE NOTE 1.

BILL OF MATERIALS

1. PACKING GLAND ASSEMBLY
(RETRACTABLE PROBE SENSOR)

2. TEE FITTING.
SEE NOTE 3

3. NEMA 7 JUNCTION HOUSING

4. PVC, KAPTON ON
FIBRE–GLASS JACKETED
CONNECTION CABLE
(UP TO 1000 FEET)

5. REMOTE ELECTRONICS
HOUSING. NEMA 7 FOR HAZ-
ARDOUS AREAS NEMA
4 OTHERWISE.

NOTE 2

FLOW METER – THERMAL

NOTES:

1. MOUNT PROBES IN VERTICAL PIPES ONLY IF THE FLOW DIRECTION IS UP.
SIDE MOUNT PROBES IN HORIZONTAL PIPES SO THAT PROBE IS HORIZONTAL.

2. MAKE PROBE U–DIMENSIONS SUCH THAT DISTANCE INDICATED IN DETAIL
IS 1–1/2" MINIMUM IN PIPELINES 3" OR LARGER. FOR PIPELINES
2" & SMALLER, PROBE SHALL PENETRATE AT LEAST TO CENTER OF PIPELINE.

3. PROBE ASS'Y SHOWN IS FOR APPLICATIONS IN 1" & LARGER PIPELINES.
FOR SMALLER PIPELINES SUPPLY AN INJECTION TUBE ASSEMBLY IN PLACE OF
THE TEE & BUSHINGS.

Figure 3.19. Thermal flow meter installation.

Maintenance and Calibration

Thermal mass flow meter probes are factory-calibrated and normally do not require calibration.

Task	Frequency
1. Check the electrical calibration of the transmitter	If the accuracy becomes questionable
2. Remove the probe from the process and examine for visible buildup or deposit of material	Every six months

Deficiencies

The following are problems with thermal mass flow meter applications.

- Wrong range.
- Water droplets forming on probe.
- Probe not positioned properly in pipe.
- Insufficient straight-run piping provided.

BIBLIOGRAPHY

Orifice Plate

1. Liptak, B. G., and K. Venczel. *Instrument Engineers Handbook of Process Measurement*, rev. ed. (Radnor, PA: Chilton Book Company, 1982).
2. "Fluid Meters, Their Theory and Application. Report of ASME Research Committee on Fluid Meters," 6th ed. (New York: American Society of Mechanical Engineers, 1971).
3. Technical Bulletins 6–110, 7–110, 7–251. Foxboro, MA: The Foxboro Company.
4. "Flange Mounted Sharp Edged Orifice Plates for Flow Measurement." ISA–RP3.2–1978 (Research Triangle Park, NC: Instrument Society of America).
5. "Manual on Installation of Refinery Instruments and Control Systems." Part 1–Process Instrumentation and Control. Section 1–Flow, 3rd ed., API RP550, (Washington, DC: American Petroleum Institute, 1977).
6. Henson, J. E., Process Instrumentation Manifolds. (Research Triangle Park, NC: Instrument Society of America, 1981).
7. Spink, K. L., *Principles and Practice of Flow Meter Engineering*, 9th ed. (Foxboro, MA: The Foxboro Company, 1967).

8. Cusick, C. F. *Flow Meter Engineering Handbook*, 3rd ed., (Fort Washington, PA: Honeywell, Inc., 1977).
9. Sprenkle, R. E. "Piping Arrangements For Acceptable Flow Meter Accuracy," *ASME Trans.* 65:345,(1945).
10. Starret, P. S., Halfpenny, P. F. and Noltage, H. B. "Survey of Information Concerning the Effects of Nonstandard Approach Conditions Upon Orifice and Venturi Meters," paper presented at Annual Winter Meeting, American Society of Mechanical Engineers, New York, NY, 1965.

Venturi Tubes and Flow Tubes

1. Liptak, B. G., and K. Venczel. *Instrument Engineers Handbook of Process Measurement*, rev. ed. (Radnor, PA: Chilton Book Company, 1982).
2. Spink, K. L., *Principle and Practice of Flow Meter Engineering*, 9th ed., (Foxboro, MA: The Foxboro Company, 1967).
3. "Instrumentation in Wastewater Treatment Plants," WPCF Manual of Practice No. 21 (Washington, DC: Water Pollution Control Federation, 1978).
4. "Fluid Meters, Their Theory and Application," Report of ASME Research Committee on Fluid Meters (New York: American Society of Mechanical Engineers, 1971).
5. "Manual on Installation of Refinery Instruments and Control Systems." Part 1–0Process Instrumentation and Control. Section 1–0Flow. API RP550, (Washington, DC: American Petroleum Institute, 1965).
6. Henson, J.E., Process Instrumentation Manifolds. (Research Triangle Park, NC: Instrument Society of America, 1981).

Averaging Pitot Tubes

1. Liptak, B. G., and K. Venczel. *Instrument Engineers Handbook of Process Measurement*, rev. ed. (Radnor, PA: Chilton Book Company, 1982).
2. Considine, D. M., *Process Instruments and Controls Handbook*. (New York: McGraw-Hill Book Company, 1974).

Turbine Flow Meters

1. Liptak, B. G., and K. Venczel. *Instrument Engineers Handbook of Process Measurement*, rev. ed. (Radnor, PA: Chilton Book Company, 1982).
2. Technical Bulletin No. T1, 16–6a (Foxboro PA: The Foxboro Company, January 1971).

Thermal Mass Flow Meters

1. Instruction and Operations Manual No. 003145, Revision N/C (San Marcos, CA: Fluid Components,Inc., September 18, 1985).

Chapter 4

Flow Measurement, Open Channel

Flumes and weirs are the most commonly used flow meters to measure open channel flow. Other meters include the Kennison nozzle and velocity-area type. This chapter covers weirs, the Parshall flume, the Palmer-Bowlus flume, the Kennison nozzle and velocity-area type open channel flow meters.

WEIRS

Application

A weir is used to measure flow in open channels where the water is relatively free of suspended solids. Weirs are suitable for metering flow under the following general conditions:

1. Flow stream should have less than 50 mg/L suspended solids.
2. Sufficient hydraulic head exists so a weir can be used. Typically, the head loss of a rectangular weir is four times that of a Parshall flume of equal size at the same flow.
3. Flow rates vary over a large range. A range of flows of 20:1 can be

119

Table 4.1. Weir Application Guidelines

Recommended	Not Recommended
Raw water	Raw sewage
Finished water	Mixed liquor
Secondary effluent	Sludge
Primary effluent	
(with provisions for	
sluicing)	

tolerated by most weirs. For weirs larger than 2.5 m (8 ft), ranges of 75 to 1 are reported. These wide ranges are not recommended.

4. The approach conditions insure that at all flow rates the flow is tranquil, free of eddies or surface disturbance. Under maximum flow, approach velocities in the upstream channel should not exceed 10 cm/s (4 in./s).

Weir application guidelines can be found in Table 4.1.

Principle of Operation

A weir is a dam or bulkhead placed across an open channel with an opening on the top through which the measured liquid flows. The opening is called the weir notch. Its bottom edge is called the crest. Normally the notch is cut from a metal plate and attached to the upstream side of the bulkhead. This is done to prevent the water from contacting the bulkhead and is known as a sharp-crested weir (see Figure 4.1). A weir without a plate where the water contacts the bulkhead is known as a broad-crested weir. This section concentrates on sharp-crested weirs.

The water depth measured at a prescribed distance upstream can be used to determine the discharge through the weir. Characteristic head versus flow relationships are governed by the weir geometry. All level measurements are made relative to the crest elevation.

The weir openings are normally fabricated in rectangular, trapezoidal, or V-notch shapes. A trapezoidal weir with a side slope of 4:1 is known as a Cipolletti weir. Figure 4.2 illustrates the weir shapes. Flow as a function of upstream head (h) is expressed by empirical equations (see Figure 4.1). General equations for each weir shape are given below. These are covered in more detail in Bibliography numbers 1 and 2.

The following equations show the relationship between flow and the measured head.

1. For a rectangular weir:
$$Q = K(L-0.2H)H^{3/2}$$

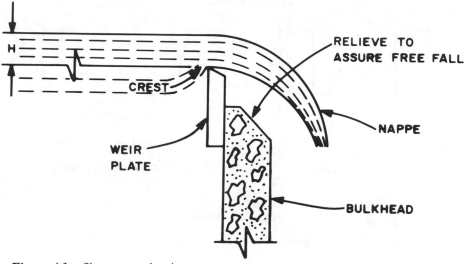

Figure 4.1. Sharp-crested weir.

2. For a Cipolletti weir:
$$Q = KLH^{3/2}$$
3. For a V-notch weir:
$$Q = K \tan 0.5 \, \theta \, H^{5/2}$$

Where   Q = Rate of flow
           L = Crest length
           H = Head of flowing liquid
           θ = V-notch angle in degrees
           K = Constant dependent the units of flow

A special rectangular weir without end contractions can be installed with the sidewalls of the channel forming the ends of the weir. This is known as a suppressed weir and is shown in Figure 4.3. When this type weir is applied, an air vent must be installed to allow free access of air beneath the nappe for free flow.

V-notch weirs are suitable for flows up to 17 kL/min (4500 gpm). Rectangular and Cipolletti weirs are capable of measuring much higher flows than the V-notch weir.

Accuracy and Repeatability

         Accuracy   ±2% for the head versus flow
                                  relationship
                          ±5% flow measurement accuracy

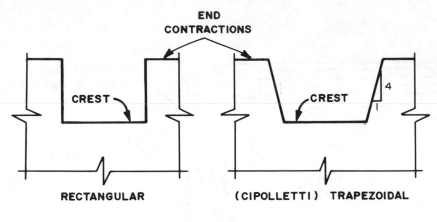

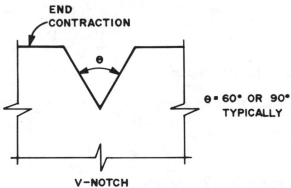

Figure 4.2.   Weir shapes.

The weir is a primary element. The accuracy of flow indicated or recorded flow is also dependent on the secondary elements, the level sensor and flow converter. Refer to Chapter 5 for more information on selecting and installing level instrumentation.

To obtain the best accuracy possible when designing and installing a weir, observe the following:

1. The minimum head (see Figure 4.1) should be 6 cm (2.5 in.) or greater.
2. The maximum head should be less than one-half the height of the weir.
3. When using rectangular or Cipolletti weirs, the maximum head is less than one-half the crest length of the weir.
4. Rectangular and Cipolletti weir crests should be level.
5. Use a V-notch weir for low-flow measurement.

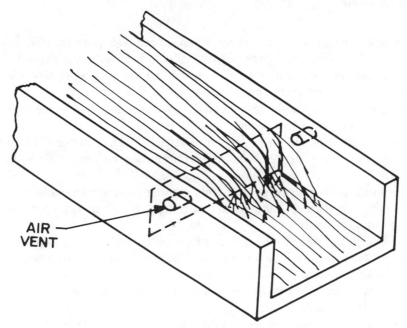

Figure 4.3. Suppressed rectangular weir.

6. All edges and corners of the weir must be sharp.
7. Weir edges should be straight, smooth, and free of burrs.
8. The approach channel should be straight and of uniform cross section for a length equal to at least 15 times the maximum head on the weir.
9. The channel should have a free fall of 15 cm (6 in.) downstream of the weir.

Manufacturers' Options

The weir is normally fabricated for each installation. Nonetheless, it may be desirable to consider the following optional features when fabricating a weir:

1. Level sensor stilling well
2. Depth gauge referenced to the crest elevation
3. A sluicing slit with cover located at the bottom of the bulkhead for flushing out solids that may collect behind the weir

Installation

- Make the upstream face of the bulkhead and weir plate smooth, and install it in a vertical plane perpendicular to the axis of the channel.
- Insure the crest is level for rectangular and Cipolletti weirs. For a V-notch weir, insure the bisecting line of the V is vertical.
- Cut the V-notch weir angle precisely and mount the plate so the angle is bisected by a vertical line.
- Machine or file the weir edges to be straight and free of burrs. Chamfer the trailing edge to obtain a crest thickness of 1−2 mm (0.03−0.08 in.).
- Install the weir so the distance from the weir crest to the bottom of the approach channel is the greater of 30 cm (12 in.) or two times the maximum head.
- Design the weir so the end contractions on each side (except suppressed weirs) will be a minimum of 30 cm (12 in.) or two times the maximum head.
- Provide air vents under the nappe on both sides of a suppressed rectangular weir.
- Position a rectangular weir so that sides are straight up and down.
- Slope the side of a Cipolletti weir outward 1 horizontal to 4 vertical.
- Make the crest length of rectangular and Cipolletti weirs at least three times the maximum upstream head.
- Construct the bulkhead opening approximately 8 cm (3 in.) larger on all sides than the weir notch.
- Slope the top of the bulkhead down to assure that the nappe falls free without hitting the bulkhead.
- Locate the level sensor next to the sidewall so it can be reached easily.
- Position the level sensor upstream of the weir at least four times the maximum head to avoid the effect of the drawdown.
- Install the depth gauge and level sensor so the zero reference elevation is the same as the weir crest elevation.

Designer Checklist

If you can answer "yes" to the following questions when designing or reviewing weir applications, the application should be correct.

- Is the level sensor located upstream at least four times the maximum weir head?

- Is the maximum downstream liquid level at least 6 cm (2.5 in.) below the elevation of the crest?
- Is the cross-sectional area of the approach channel at least eight times the cross section of the water overflowing the crest at maximum flow?
- Is the approach channel straight and of uniform cross section for a length at least 15 times the maximum head?
- If a suppressed rectangular weir is being used, has an air vent been provided under the nappe on both sides of the channel?
- At minimum flow, does the head above the crest exceed 6 cm (2.5 in.)?
- Is the weir notch sized and shaped so the nappe will clear the bulkhead and fall free?
- Is the length of the bulkhead end contraction on each side of the weir opening at least two times the maximum head above the crest or 30 cm (12 in.), whichever is larger?
- Is the height of the weir crest above the channel bottom greater than twice the maximum head or 30 cm (12 in.), whichever is larger?
- Is the length of the rectangular or Cipolletti weir crest at least three times the maximum head?
- Is the maximum velocity in the approach channel less than 10 cm/s (4 in./s)?
- Is a depth gauge installed so periodic checks can be made on the level sensor?

Maintenance and Calibration

A depth gauge mounted adjacent to the level sensor is recommended. Make periodic level readings and determine the flow by calculation or from a lookup table. Compare remote flow readings with the value determined from the visual inspection as a conformance check on the calibration of the level sensor and flow converter.

Task	Frequency
1. Check level sensor with other level or flow indicators to determine if calibration of the secondary system is required	Weekly
2. Remove accumulation of bottom deposits as required	Weekly
3. Clean stilling well (if used) for solids accumulation	Weekly

4. Relevel weir, check all   Annually
   connections
5. Check reference zero on   Annually
   the depth gauge
6. Check for leaks around the   Annually
   weir
7. Inspect weir notch for   Annually
   nicks, dents, and rounding
   of upstream corners

Deficiencies

The following problems have been encountered in existing weir installation.

- Insufficient head during low flow conditions so there is no free air space under the nappe.
- Suppressed rectangular weirs without air vents under the nappe.
- Insufficient relief on the bulkhead so the nappe strikes the bulkhead, interfering with the free fall.
- Pool level downstream is too high so insufficient free fall exists.
- Weir notches cut from metal plate stock and installed without the edges finished to proper thicknesses, shape, or straightness.
- Rectangular or Cipolletti weirs installed without leveling the crest.
- Incorrectly cut angles and V-notch weirs.
- Level sensors located too close to the weir.
- Level sensors or gauges not zero referenced to the bottom of the weir notch.

PARSHALL FLUME

Applications

Parshall flumes are the most common method for metering flow under the following general conditions:

1. Flow is in an open channel.
2. Hydraulic head loss must be minimized. Typically, head loss for a Parshall flume is approximately 25% of the for a weir of equal capacity.
3. Sediment or solids in the measured stream (velocities in the flume tend to scour and flush away deposits).
4. Anticipated flow rates will vary widely. Depending on flume size and

the accuracy of level measurement, a maximum to minimum flow range of 20:1 is reported for Parshall flumes.

5. Approach conditions upstream of the flume will insure that the entering flow is tranquil and uniformly distributed.

Parshall flume application guidelines can be found in Table 4.2.

Principle of Operation

The Parshall flume is a device for measuring liquid flow in open channels. The Parshall flume is a constriction of the channel that develops a hydraulic head which is proportional to flow. Figure 4.4 illustrates the shape and sections of a Parshall flume. Parshall flume sizes refer to the width of the throat section. Flumes are available in sizes from 0.025 m (1 in.) up to 15 m (50 ft). Large flumes are constructed on site, but smaller flumes can be purchased as prefabricated structures or as lightweight shells which are set in concrete. Dimensions for the fabrication of Parshall flumes are contained in the *Water Measurement Manual* published by the United States Department of the Interior.

If a Parshall flume has been constructed to standard dimensions and properly set, it is possible to calculate flow through the flume by measuring level at a single point. The location for the level measurement is shown as H_a in Figure 4.5. Flow is approximately proportional to the three halves power of the hydraulic head. Simplified equations can be found in Bibliography number 2.

Two flow conditions can exist in the Parshall flume: free flow and submerged flow. Free flow exists when the only restriction is the throat width and the water is not slowed by downstream conditions. If the flow through the flume increases sufficiently, the downstream channel level may rise and impede the discharge from the flume, thereby slowing the fluid velocity. This is known as submerged flow.

It might be expected that the flume discharge would be reduced as soon as the tailwater level exceeds the elevation of the crest. Tests have shown that this is not the case. Free flow conditions can still exist even with some degree

Table 4.2. Parshall Flume Application Guidelines

Recommended	Not Recommended
Raw water	Sludges
Finished water	Chemicals
Filtered water	
Raw sewage	
Primary effluent	
Secondary effluent	
Plant final effluent	
Mixed liquor	

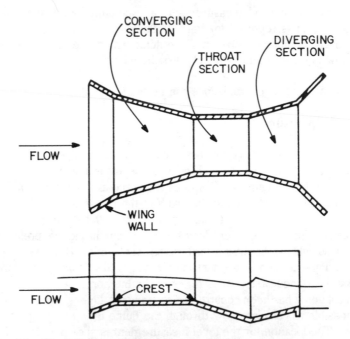

Figure 4.4. Parshall flume flow element.

of submergence. Figure 4.6 shows the head relationship for free flow. Table 4.3 lists flume sizes and the limits of free flow submergence. For free flow conditions, the depth measurement (H_a) can be used to calculate the flume discharge flow.

Although Parshall flumes can operate with a submergence greater than those shown in Table 4.3, a second level measurement and a correction factor is required to calculate flow. There is a loss of accuracy at submergence, and it is not a recommended design.

Two elements are involved in obtaining a flow measurement with a Parshall flume. The primary element is the Parshall flume structure and the secondary element is the level measuring device. For additional information on methods to measure level refer to Chapter 5.

Accuracy and Repeatability

Accuracy	± 3% of flow for the depth discharge equation
	± 5% of flow for the combined flume and level measurement
Repeatability	± 0.5% of flow

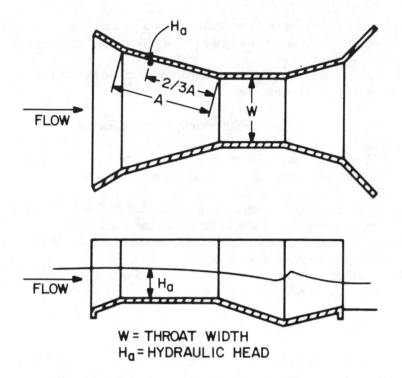

W = THROAT WIDTH
H_a = HYDRAULIC HEAD

Figure 4.5.   Head/width parameters.

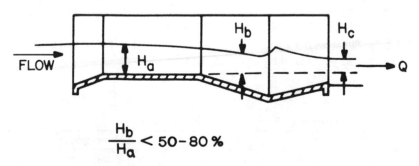

$$\frac{H_b}{H_a} < 50\text{-}80\,\%$$

Figure 4.6.   Free flow submergence.

The accuracy of a measurement derived using a Parshall flume depends on the combination of the accuracies of the primary (flume) and secondary (level measurement) elements.

Additional sources of error, if uncorrected, can decrease the accuracy of the flow measurement. These include:

1. Deviations of the throat width from standard dimensions.
2. Longitudinal slope of the floor in the converging section. Tests on a 0.75 m (3 in.) flume demonstrated that a downward sloping floor produced added errors of 3–10% from low to high flow conditions.
3. Transverse slope of the flume floor.
4. Approach conditions which do not produce a smooth flow with uniform velocity distribution parallel to the center line of the flume.
5. Incorrect zero reference of the level measurement device to the center line elevation of the crest.
6. If a stilling well is used, the connector hole is improperly sized.
7. Incorrect zero reference of the level measurement device.

Manufacturers' Options

Although Parshall flumes can be constructed on site, most flumes are prefabricated structures or liners for setting in concrete. Some available options include:

- Material of construction
- Stilling well connection
- Attached stilling well
- Depth gauge integrally mounted in the converging section sidewall
- A cavity for a characterized capacitance level measuring probe molded into the flume sidewall
- Removable bubbler tube installed in the sidewall
- A large flume with a smaller one mounted internally. The smaller flume is removed when flow exceeds its capacity.

Table 4.3. Submergence Limits

Flume Size	% Submergence
0.025, 0.05, 0.075 m (1, 2, 3 in.)	50
0.15, 0.23 m (6, 9 in.)	60
0.3–2.4 m (1-8 ft)	70
3–15 m (10-50 ft)	80

Installation

- Construct or install the flume so that the floor section is level longitudinally and transversely.
- Establish the flume floor's elevation to prevent submergence conditions at maximum flow.
- Plan the flume installation to allow access for inspection of the flume to ensure correct elevation and leveling of the floor.
- Provide an approach channel long enough to create a symmetrical, uniform velocity distribution and a tranquil water surface at the flume entrance. Ideally, provide 10 channel widths of straight run upstream of the flume inlet. As a minimum, provide 10 throat widths upstream.

Designer Checklist

If you can answer "yes" to the following questions when designing or reviewing Parshall flume applications, the application should be correct.

- Has the smallest practical Parshall flume been selected for the anticipated range of flows?
  a. as a guide, the maximum flow expected should fall within 70 to 100% of the maximum capacity for the selected flume size (1, 3).
  b. a depth of at least 0.15 m (0.5 in.) should exist at the minimum actual flow.
- Is the flume floor elevation high enough (relative to downstream conditions) to prevent submerged flow?
- Does a sufficient straight run of pipe or channel exist upstream?
- Will the upstream flow be nonturbulent and wave-free?
- Do any downstream obstructions exist which could cause a restriction to the discharge of the flume?
- Is a depth gauge included for calibrating the flume?
- Is the level sensor correctly located on the flume?
- Is the level sensor zero correctly referenced?
- If required for the application, is a stilling well provided?
- The following items pertain to the stilling well:
  a. is the vertical height extended below and above the anticipated operating depths in the flume?
  b. is the flume opening for the stilling well sized large enough to avoid sensor lags and plugging?
  c. has the flume opening been correctly located on the length of the converging section wall?
  d. is the flume opening positioned below the lowest flow operating level?

e. if the flume is used to measure raw sewage or mixed liquor flow, has a water purge been piped to the stilling well?

Maintenance and Calibration

A depth gauge mounted on the converging section is recommended so manual readings and flow calculations can be made to check remote flow indicators or recorders.

Parshall Flume

Task	Frequency
1. Check the depth gauge with other level or flow indications for the flume to determine if calibration of the secondary system is required	Weekly
2. Wipe down the flume walls to remove slime or other buildup	Monthly or as necessary
3. Remove bottom deposits as required	Every 3 months
4. Check the zero of the reference depth gauge	Every 3 months
5. Examine the flume surfaces for signs of deterioration and wear	Every 6 months

Stilling Well (if used)

Task	Frequency
1. Remove solids accumulation	Every 3 months or as necessary
2. Check and adjust water purge, if used	Daily

Deficiencies

The following problems are encountered in existing Parshall flume installations.

- Insufficient straight channel upstream, resulting in nonuniform velocity distribution through the throat.

- Sloping floor in the converging section.
- Low flume floor elevation (relative to downstream channel level) resulting in submerged flow condition.
- Measuring the depth of the hydraulic head at the wrong location.
- Using an incorrect equation for calculating flow from level.
- No provisions for purging the stilling well when measuring a solids-bearing stream (buildup of deposits in the stilling well renders the level sensor inoperable).
- The foundation is not water tight, allowing leakage under or around the flume.
- Flume sized too large for the operating flow range.

PALMER-BOWLUS FLUME

Application

Palmer-Bowlus flumes are normally installed in sewers between sections of pipe. The following general conditions apply when considering a Palmer-Bowlus flume.

1. Open channel with round bottom or partially filled pipes (less than 90% filled) where fabrication of a flow transition approach section to accommodate a Parshall flume is not practical.
2. Hydraulic head loss must be minimized.
3. Sediment or solids in the measured stream (velocities in the flume tend to flush away deposits).
4. Variations in the flow rates are expected to be within a 10:1 range.
5. The flow entering the flume is subcritical (velocity is less than in the flume throat), nonturbulent, and uniformly distributed.

Table 4.4 shows application guidelines for Palmer-Bowlus flumes.

Table 4.4. Palmer-Bowlus Flume Application Guidelines

Recommended	Not Recommended
Sewage	Sludges
Water	Chemicals

Principle of Operation

The Palmer-Bowlus flume is a restriction in the channel which produces critical flow through the throat of the flume. This restriction also causes the water to back up upstream of the flume. Figure 4.7 shows the sections of a Palmer-Bowlus flume. The throat cross section is trapezoidal.

Flow rate is related to upstream depth. This relationship is derived analytically from an energy balance between the point of depth measurement and the flume throat. The point of depth measurement is about 1/2 pipe diameter upstream from the entrance to the flume (refer to Figure 4.8).

Palmer-Bowlus flumes are subject to free and submerged discharge conditions. Free discharge will prevail as long as the ratio of downstream to measured depth (H_D/H_S) does not exceed 0.90. One reference suggests that the H_D/H_S ratio should not exceed 0.85. When the H_D/H_S exceeds 0.90, submerged flow exists. Correction factors are not available for submerged flow conditions.

The Palmer-Bowlus flume size is determined by its cross-section diameter. Prefabricated flume liners are available in sizes from 0.1 to 1.5 m (4 to 60 in.). To avoid selecting an oversized flume, care should be taken to base flume size on actual flow rather than nominal pipe size.

Two elements are required to measure flow with a Palmer-Bowlus flume. The primary element is the Palmer-Bowlus flume structure. The secondary element is a level measuring device. For additional information on equipment to measure level, refer to Chapter 5.

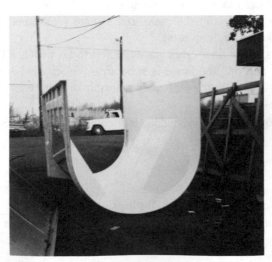

Figure 4.7. Palmer-Bowlus flume.

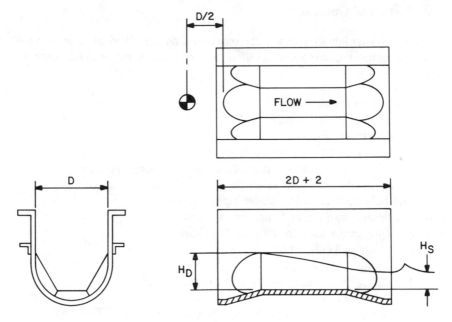

Figure 4.8. Free flow/depth relation.

Accuracy and Repeatability

Accuracy
$\pm 3\%$ of full scale for the flume primary element
± 5 to $\pm 10\%$ of full scale for flume and a level measuring device

A given flow differential will produce a relatively small head differential thus requiring a sensitive, accurate level measurement for best results.

There are additional sources of error which can add to the inaccuracy of the meter. The principal ones are:

1. Longitudinal slope of the flume floor greater than 1.5%.
2. Any transverse slope of the flume floor.
3. Approach conditions which do not produce a smooth flow with uniform velocity distribution parallel to the center line of the flume.
4. Incorrect zero reference of the level measurement device.
5. Where stilling wells are used, connector hole is improperly sized.

Manufacturers' Options

Palmer-Bowlus flumes typically are purchased as a prefabricated liner to be set in concrete or grouted into a half-section of pipe. Some available options include:

1. Material of construction
2. Alternative configurations:
   a. basic insert flume
   b. flume with integral approach section
   c. cutback flume for insertion into a manhole discharge pipe
3. Flanged ends
4. End bulkheads to fit in a larger pipe
5. For flumes with integral approach sections:
   a. depth gauge flush mounted in the sidewall
   b. stilling well connection
   c. attached stilling well
   d. removable bubbler tube installed in the sidewall
6. Nested flumes
   A large flume with a smaller one mounted internally. The smaller flume is removed when flow exceeds its capacity.

Installation

- Install the flume so that the floor of the flume is level longitudinally and laterally.
- Establish the flume floor elevation to prevent submergence conditions at maximum flow.
- Plan the flume installation to allow access for inspection to ensure correct elevation and leveling of the floor.
- Use an approach channel long enough to create a symmetrical, uniform velocity distribution and a tranquil water surface at the flume entrance. A general rule is the greater of either 20 throat widths or 8 pipe diameters of straight run should exist upstream of the flume inlet.

Designer Checklist

If you can answer "yes" to the following questions when designing or reviewing Palmer-Bowlus flume applications, the application should be correct.

- Has the smallest practical size Palmer-Bowlus flume been selected for the anticipated range of flows?
   a. as a rule of thumb, the flume throat width is one-third to one-half of the pipe diameter.

    b. another sizing guide is that the maximum flow expected should fall within 70 to 100% of the maximum capacity for the selected flume size (1, 3).

    c. a minimum depth of 0.15 m (0.5 in.) should exist at the minimum actual flow.

- Is the flume floor elevation sufficient to avert submerged flow conditions?
- Does a straight channel longer than 20 throat widths or 8 pipe diameters exist upstream of the flume?
- Will the anticipated upstream flow provide nonturbulent, wave-free approach conditions?
- Is the upstream approach velocity subcritical?
- Do any downstream obstructions exist which could restrict the discharge of the flume?
- Is a reference gauge provided for measuring depth in the flume?
- Is the point of level measurement correctly located on the flume?
- Is level sensor zero-referenced to the floor of flume at the center line of the throat?
- If required, is a stilling well provided for level measurement?

The following items pertain to the stilling well:

- Does the vertical height extend below and above the anticipated operating depths in the flume?
- Is the flume opening for the stilling well sized large enough to avoid sensor lags and plugging?
- Has the flume opening for the stilling well been correctly located on the length of the converging section wall?
- Is the flume opening for the stilling well positioned below the operating level at minimum flow?
- Is a fresh water purge piped into the stilling well?

Maintenance and Calibration

It is recommended that a depth gauge be mounted upstream of the flume entrance so that manual readings and flow calculations can be made to check remote flow indicators or recorders.

Palmer-Bowlus Flume

Task	Frequency
1. Check the depth gauge with other level or flow indications for the flume to determine if calibration of the secondary system is required	Weekly
2. Wipe down the flume walls to remove slime or other buildup	Monthly or as necessary
3. Remove bottom deposits as required	Every three months
4. Check the zero of the reference depth gauge	Every three months
5. Examine the flume surfaces for signs of deterioration and wear	Every six months

Stilling Well (if used)

Task	Frequency
1. Remove solids accumulation	Every three months or as necessary
2. Check and adjust water purge, if used	Daily

Deficiencies

The following problems are encountered in Palmer-Bowlus flume installations.

- Flume sized too large or small for the operating flow range.
- Insufficient straight channel upstream resulting in nonuniform velocity distribution through the throat.
- Excessive transverse or longitudinal slope to the flume floor.
- Measuring the depth of the hydraulic head at the wrong location.
- Level sensor zero and the flume floor elevation not equal.
- Using an incorrect equation for calculating flow from level.
- No provisions for purging the stilling well when measuring a solids-bearing stream (buildup of deposits in the stilling well renders the level sensor inoperable).

- The foundation of the flume is not water tight, allowing leakage under or around the flume.
- Difficult to zero level instrument due to shape of flume.

KENNISON NOZZLE

Application

The Kennison nozzle is a proprietary product designed to measure flow in partially filled pipes. Kennison nozzles are designed to flush through solids without accumulation and are generally considered to have a maximum to minimum measurement ratio of 10:1. A Kennison nozzle could be considered for flow measurement only when all of the following general conditions apply:

1. Pipe size is between 0.15 m (6 in.) and 0.9 m (36 in.).
2. The pipe does not flow full.
3. High and low operating flows fall respectively within the 100% and 10% capacity of the desired nozzle size.
4. The level of the liquid downstream will always be below the bottom of the nozzle so that free discharge exists.

Kennison nozzle application guidelines are presented in Table 4.5.

Principle of Operation

The Kennison nozzle is a constriction to liquid flow of known geometry which will produce a hydraulic head at the area of the constriction. When the nozzle is operating above minimum flow rate (10% of the maximum nozzle flow) the head is essentially linear to flow rate, providing that free discharge exists. Figures 4.9 and 4.10 illustrate a Kennison nozzle and the relationship between flow rate and head.

The Kennison nozzle is designed to measure flow in partially filled pipes. If high flow conditions fill the pipe, the nozzle will overflow. Any measurements taken during this period will be inaccurate.

For low flow and low head applications, the manufacturer has available a half section Kennison nozzle.

Table 4.5.   Kennison Nozzle Application Guidelines

Recommended	Not Recommended
Raw sewage	Filled lines
Partially filled lines	
Raw water	

Figure 4.9.   Kennison nozzle.

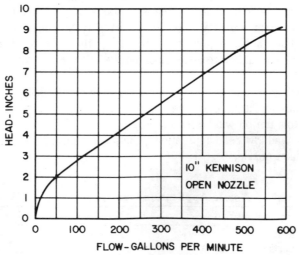

Figure 4.10.   Typical rating curve for a 10 in. Kennison nozzle.

Accuracy and Repeatability

Accuracy $\pm 1\%$ of actual for flow between 10% and 100% of capacity
± 3 to $\pm 5\%$ of actual for flows less than 10% of maximum

The accuracy is dependent on the combined accuracy of the nozzle (primary element) and the level measuring device (secondary element). The manufacturer claims a primary element accuracy of $\pm 0.1\%$ if the nozzle is flow-calibrated at a hydraulics laboratory.

Manufactured Options

The Kennison nozzle is a proprietary product of BIF, a unit of General Signal. Available options include:

1. A half section nozzle for measuring flow under low hydraulic head conditions, or lower than standard flow rates
2. Cast iron or fiberglass material of construction
3. Pressure tap connection on left or right side of nozzle
4. Hydraulically operated vent cleaner for remotely or automatically initiated cleaning of the pressure tap
5. Flow indicator
6. Stilling well, level float, and transmitter with an output characterized for the nozzle
7. Upstream spool piece and capacitance probe linearized to the nozzle

Installation

- Elevate the nozzle invert (bottom) above the level of the downstream surface to ensure free discharge conditions.
- Level the nozzle in the horizontal plane (lengthwise and crosswise).
- Install a thick gasket between the nozzle flange and pipe to facilitate leveling.
- Ensure a uniformly distributed nonturbulent flow by following the manufacturer's recommended approach conditions. They are:
  a. eight pipe diameters of unobstructed straight pipe upstream of the nozzle
  b. slope of the pipe line should not exceed the tabulated limits for the nozzle size selected as indicated in Table 4.6
  c. line velocity should not exceed tabulated approach velocities for the nozzle size selected as indicated in Table 4.7

- Set the invert of the nozzle flange and the mating pipe flange at the same elevation.
- Consult the manufacturer's engineering data bulletin for details on the installation of a secondary level measuring system.

Designer Checklist

If you can answer "yes" to the following questions when designing or reviewing Kennison nozzle applications, the application should be correct.

- Have the manufacturer's rating curves been consulted to ensure the selected nozzle size meets the following criteria?
- Has the smallest practical size been selected for the anticipated range of flows?
- Is minimum anticipated flow greater than 10% of nozzle capacity?
- Will the pipe be less than full under maximum flow conditions?
- Is the invert of the nozzle above the downstream liquid level at all times?
- Is the straight pipe upstream of the nozzle greater than 8 diameters?

Table 4.6. Limiting Slope for Approach Piping

Nozzle Size (In.)	Slope
6	0.0070
8	0.0050
10	0.0040
12	0.0033
16	0.0027
20	0.0023
24	0.0021
30	0.0020
36	0.0020

Table 4.7. Maximum Line Velocity for Kennison Nozzle Installation

Nozzle Size (In.)	Maximum Line Velocity (m/sec)
6	0.67
8	0.67
10	0.70
12	0.79
16	0.91
20	1.00
24	1.13
30	1.13
36	1.13

- Is slope of the pipe less than the manufacturer's limits?
- Is approach velocity within the limits specified by the manufacturer?

Maintenance and Calibration

Install either a pressure gauge or manometer on the zero check tap. Use readings from this indicator to pick the flow from the manufacturer's rating curve.

Task	Frequency
1. Operate the vent cleaner to remove any buildup or obstruction in the pressure tap	Weekly
2. Measure the hydraulic head manually to determine if calibration of the secondary system is required	Weekly
3. Inspect the secondary level system	Weekly
4. Remove solids buildup as required	Every six months

Deficiencies

Improper approach conditions constitute the most common problem in the use of the Kennison nozzle. Either the slope of the approach piping is too great or the approach velocity is too high.

VELOCITY-AREA

Applications

Velocity-area meters are available in three different types: magnetic probe, transmissive sonic, and doppler.

Velocity-area flow meters are suitable for the application under the following conditions:

1. You desire to measure flow in existing sewers or aquaducts and do not want to modify the piping.
2. You do not know the slope or roughness coefficient accurately enough

to use the Manning equation to predict flow from depth measurement only.

3. Flow reversals are possible.
4. You want a portable open channel flow measuring device.
5. The liquid conductivity is greater than 5 μmhos/cm if you use the magnetic-probe type meter.

Principle of Operation

Velocity-area flow meters use velocity, depth, and channel geometry information to determine the flow.

Velocity is determined using transmissive sonic sensors, Doppler sensors, or insertion magnetic probes. These meters have the same principle of operation as described in Chapter 2.

Level is measured using ultrasonic level sensors or piezoelectric pressure sensors. See Chapter 5 for a discussion of ultrasonic level sensors and Chapter 6 for a discussion of pressure sensors.

Accuracy and Repeatability

Accuracy	$\pm 2\%$ of full scale velocity
	$\pm 0.5\%$ of full scale for level
	$\pm 5\%$ of full scale for flow

Accuracy can be significantly affected at low flow rates if the velocity profile changes significantly. If there is reason to suspect a bad velocity profile, take readings at the 0.2 and 0.8 depth from the surface. These readings will approximate the average velocity.

Manufacturers' Options

1. Transmitter:
   a. totalizer
   b. 4–20 mAdc or flow proportional pulse output
   c. plug in modules for various channel geometries
2. Power:
   a. 115 Vac, 50–60 Hz
   b. 220 Vac, 50–60 Hz
   c. 24 Vdc for portable units

Installation

Figure 4.11 shows two mounting options for a transmissive type velocity-area meter. Figure 4.12 shows the normal mounting option for magnetic probe-type velocity-area meters. If silt or grit settling is expected, the probe can be repositioned. The level sensing calibration must be adjusted to account for this.

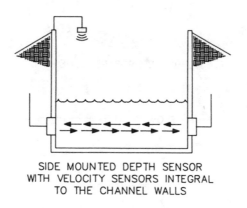

SIDE MOUNTED DEPTH SENSOR
WITH VELOCITY SENSORS INTEGRAL
TO THE CHANNEL WALLS

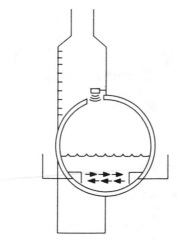

MANHOLE MOUNTING TECHNIQUE

Figure 4.11.   Transmissive velocity-area meter mounting.

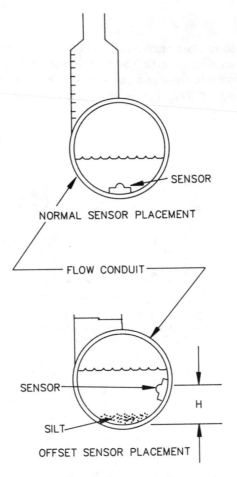

Figure 4.12.   Insertion magnetic velocity-area probe mounting.

- Provide 10 diameters straight run upstream and 5 diameters down-stream of meter.
- Place the meter in an area that is easy to inspect.
- Avoid locations with excessive flow turbulence. The fluid surface should appear smooth and undisturbed.
- Carefully measure the channel at the meter location to accurately determine channel depth/area relationships.
- If silt or grit is expected, locate the sensor head above the grit line.

Designer Checklist

If you can answer "yes" to the following questions when designing or reviewing velocity-area flow meter applications, the application should be correct.

- Is the fluid conductivity greater than 5 μmhos/cm for insertion type magnetic velocity-area meters?
- Are the upstream and downstream flow conditions met?
- Have you provided convenient access for maintenance?

Maintenance and Calibration

Task	Frequency
1. Remove silt, grit or debris	Weekly or as required
2. Inspect probes for movement or damage	Every three months
3. Calibrate transmitter	Monthly

Deficiencies

The following problems have been encountered with velocity-area flow meters:

- Algae or grease buildup on insertion type meters affects accuracy.
- Velocity measurement is not representative of actual under varying flow conditions.
- Transmissive velocity probes not aligned properly.
- Surcharges submerge sonic level probe.
- Silt or grit builds up around probes.
- Large debris dislodges probes. Probes inadequately protected.
- Flow profile is not fully developed.

BIBLIOGRAPHY

Weirs

1. Water Measurement Manual, 2nd ed., U.S. Department of the Interior, Bureau of Reclamation (Washington, DC: U.S. Government Printing Office, 1981).
2. Grant, D. M., *Open Channel Flow Measurement Handbook* (Lincoln, NE: ISCO, Inc., 1981).

Parshall Flume

1. Grant, D. M., *Open Channel Flow Measurement Handbook*, 2nd ed. (Lincoln, NE: ISCO, Inc., 1981).
2. Liptak, B. G., and K. Venczel. *Instrument Engineers Handbook of Process Measurement*, rev. ed. (Radnor, PA: Chilton Book Company, 1982).
3. Water Measurement Manual, 2nd ed., U.S. Department of the Interior, Bureau of Reclamation (Washington, DC: U.S. Government Printing Office, 1981).
4. Kulin, G., "Recommended Practice For the Use of Parshall Flumes and Palmer-Bowlus Flumes In Wastewater Treatment Plants." EPA–600/2–84–186 (Cincinnati, OH: U.S. Environmental Protection Agency, November, 1984).
5. "Instrumentation In Wastewater Treatment Plants." Manual of Practice No. 21 (Washington, DC: Water Pollution Control Federation, 1978).

Palmer-Bowlus Flume

1. Grant, D.M., *Open Channel Flow Measurement Manual*, 2nd ed. (Lincoln, NE: ISCO, Inc., 1981).
2. Metcalf and Eddy, Inc., *Wastewater Engineering: Treatment/Disposal/Reuse*. (New York : McGraw-Hill Book Company, 1979).
3. Kulin, G., " Recommended Practice For the Use of Parshall Flumes and Palmer-Bowlus Flumes in Wastewater Treatment Plants," EPA–600/2–84–186. (Cincinnati, OH: U.S. Environmental Protection Agency, November, 1984).
4. *Flow: Its Measurement and Control In Science and Industry, Vol. II.* (Research Triangle Park, NC: Instrument Society of America, 1981).

Kennison Nozzle

1. "Kennison Open Flow Nozzle," Engineering Data Sheet No. 135.21–1. (West Warwick, RI: BIF, a unit of General Signal).
2. Grant, D.M., *Open Channel Flow Measurement Handbook*, 2nd ed. (Lincoln, NE: ISCO, Inc., 1981).

Velocity-Area

1. Model 250 Flowmeter Technical Manual (Gaithersburg, MD: Marsh-McBirney, Inc.).

Chapter 5

Level Measurement

There are a number of different techniques used to measure level. Technologies used for level measurement include pressure, capacitance, and sonic. This chapter includes bubblers, capacitance probes, floats, and sonic and ultrasonic level sensors.

This chapter does not cover load cells and strain gauges which can be used to measure chemical levels in storage tanks. Pressure cells can also be used to measure level. Pressure instruments are discussed in Chapter 6.

BUBBLERS

Applications

Bubbler level measurement instruments are used in water and wastewater treatment processes for measuring both liquid level and differential liquid level.

Bubblers are frequently used to sense the hydraulic head created by flumes and weirs in open-channel flow measurement. A special signal converter indicates flow based on the level sensed by a bubbler. This section addresses the use of bubblers in open tanks which is applicable to open-channel flow

Table 5.1. Bubbler Level Measurement Application Guidelines

Recommended	Not Recommended
Liquid treatment processes	Digesters Volatile chemical storage tanks

measurement. Bubbler level measurement application guidelines are listed in Table 5.1.

Principle of Operation

An open-ended pipe, called the bubbler tube or dip tube, is connected to an air supply and positioned in the process so that the open end is set at a reference level. A constant air rate-of-flow regulator is used to maintain air in the tube with enough excess to continually bubble out the open end. Thus, the air pressure in the pipe is equal to the head of the process liquid above the reference level.

A pressure transmitter connected to the bubbler tube measures the pressure of the dip tube. For water, the level is equal to the pressure sensed by the transmitter. For measuring other liquids, the transmitter must be calibrated for that liquid's specific gravity. For closed tanks, a differential pressure transmitter is used, with the high-pressure port connected to the bubbler tube and the low-pressure port connected to the gas space in the top of the tank. Typical bubbler applications are shown in Figure 5.1.

Because of head loss caused by air flow in the tube and connecting pipe, pressure at the transmitter will not be exactly the same as at the open end of the bubbler tube. This difference in pressure necessitates minimizing pipe and fittings between the rate-of-flow air regulator and the dip tube.

Airflow head is affected by bubble formation. To minimize errors, the bottom of the tube usually has a notch or an angular cut to produce a continuous stream of small equally sized bubbles. Since buildup of process solids on the end of the tube will alter bubble formation, the tube end must be kept clean.

The air supply rate is controlled by a pressure regulator and a flow control valve. Typical airflow rates are 8–30 cc/s (1–4 cfh). Frequently, a purge/rotameter is used to adjust the airflow. The air supply can be from instrument air, plant air, compressed gas tanks, or dedicated bubbler compressor. For applications requiring infrequent level readings, a hand-operated pump can be used.

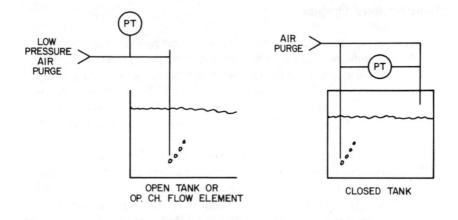

OPEN TANK OR
OP. CH. FLOW ELEMENT   CLOSED TANK

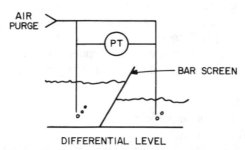

DIFFERENTIAL LEVEL

Figure 5.1. Typical bubbler applications.

Accuracy and Repeatability

Accuracy ± 0.5 to $\pm 1\%$ of full scale

The accuracy of a level measured by a bubbler system is dependent on the uncertainty of the pressure measuring device (see Chapter 6), process fluid specific gravity, head loss in the bubbler system, barometric pressure, and the temperature of both the process fluid and the bubbler system air.

Repeatability is dependent on variances from standard conditions of any of the uncertainties listed above.

Manufacturers' Options

1. A separate timer and valve control package for periodic cleaning of the bubbler tube is available. A high pressure air purge removes buildup of material at the end of the tube when the pressure measuring device is momentarily isolated from the system and full air supply pressure is applied to the bubbler tube at periodic intervals from 8 to 24 hours.
2. Bubbler systems can be furnished with a dedicated air supply consisting of:
   a. a compressor, or where extra reliability is required, two compressors with automatic failover. Intermittent duty compressors, capable of producing the high pressure purge required, range from 180–370 W (1/4 to 1/2 hp). Adequate purge pressure for most wastewater applications is about 500 kPa (60 psig).
   b. an air dryer
   c. an air filter
   d. a pressure tank with a capacity of about 0.01 m³ (2 gal)
   e. Where the oxygen contained in an air supply system is objectionable, nitrogen or another inert gas may be substituted. For explosive, volatile, or hazardous atmospheres complete intrinsic safety can be achieved by using a pneumatic pressure signal for remote indication.
3. Enclosure:
   a. NEMA 1, general purpose
   b. NEMA 4, watertight
4. Usually a thermostatically controlled heater is available.
5. Available alarms include low air flow, low air supply pressure, purge-in-progress, and compressor failure.

Installation

The bubbler tube should be rigidly supported at a convenient location in the tank. The opening of the tube is the lowest level that can be detected, so set the tube depth at or below the lowest level at which a measurement is needed. Notch the tube opening to produce a continuous flow of small bubbles.

Fabricate the bubbler tube from 1.25 cm (1/2 in.) diameter stainless steel tubing or galvanized pipe. Properly supported, this makes a rigid installation which can withstand turbulence and wave action. A tee with one branch plugged, when installed on top of the bubbler tube, provides an opening for a cleaning rod when the high pressure air purge cannot remove bubbler tip restrictions.

The bottom of the tube should be at least 8 cm (3 in.) from the tank bottom to avoid solids buildup on the tank floor. This offset must be included in the zero reference level for the liquid in the tank. An exception to this is a bubbler installed in a flume or ahead of a weir. In flume applications the bubbler tip must be at the same elevation as the flume floor or if elevated, the degree of elevation must be compensated for in the flow calculation. In weir applications the bubbler tip must be at the same elevation as the bottom of the notch, or if below the notch, the degree of offset must be compensated for in the flow calculation.

To minimize level measurement errors caused by air flow head loss, the air flow controller must be mounted as close to the dip tube as possible and connected with a minimum of fittings and tubing. For 1 cm (1/4 in.) tubing, the distance from the air purge regulator to the bubbler tube should not exceed 15 m (50 ft).

To ensure that the air purge tubing is free from traps where moisture condensate can collect, install the tubing with a continuous downward slope from the pressure transmitter and the air flow controller to the bubbler tube.

In open tanks and flumes, for periodic reference checks and to facilitate recalibration if the tube is removed for cleaning or replacement, install a depth (staff) gauge in the tank at a location visible from the dip tube. Zero on this gauge must correspond to the bubbler tube's zero reference.

A typical installation schematic is shown in Figure 5.2. Maintenance access is needed for the clean-out tee and for the bubbler system enclosure. Installation of the differential pressure transmitter is addressed in Chapter 6.

Designer Checklist

If you can answer "yes" to the following questions when designing or reviewing bubbler level meter applications, the application should be correct.

- Can air be passed through the process fluid? If not, can another gas, such as nitrogen, be substituted for air?
- Is the tank open or vented? If not, is accumulation of air acceptable?

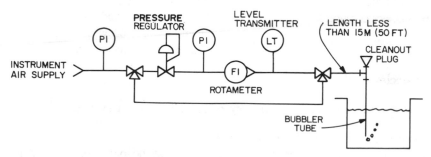

Figure 5.2.   Open tank bubbler installation.

- Are head losses from the air flow regulator to the bubble tube minimized?
- Is the purge line length from the air flow regulator to the bubbler tube less than 15 m (50 ft)?
- For process streams containing more than about 100 mg/L suspended solids, is automatic purging included?
- Does the air supply reliability match the need for level measurement reliability?
- Is the bubbler tube mounted securely?
- Is the clean-out tee accessible?
- Is the bubbler enclosure suitable for its environment?
- For open tanks, does the pressure transmitter reference the same gas space? For example, a bubbler tube mounted outside and a transmitter mounted inside a building will be exposed to different atmospheric conditions. Can this difference be tolerated?

Maintenance and Calibration

Maintaining and calibrating the differential pressure transmitter is discussed in Chapter 6.

Task	Frequency
1. Check air flow	Daily, or if the unit has a low-flow alarm detector, only at calibration
2. Purge tube	Weekly for solids bearing fluids
3. Clean tube	At calibration When fluid enters
4. Check air filter	Weekly. If the unit has a low-flow alarm detector, then check filter only at calibration.
5. Calibrate	Every two months
6. Inspect compressor	Depends on type and size of compressor. Follow the manufacturer's recommendation.

Deficiencies

The following problems are commonly reported for bubbler systems.

- Compressor failure, and measurement loss.
- Tube bubbler opening does not stay clean because no purge or cleanout is available.

- Condensate collection trapped in the purge tubing.
- Purge tubing lines too long or too small, creating excessive head losses.
- The bubbler tube tip does not correspond with the desired zero level reference.

CAPACITANCE PROBES

Applications

Capacitance probes are used to measure liquid levels throughout water and wastewater treatment plants. For this discussion, two kinds of probes are identified: capacitance and capacitance with compensation for coating. The compensated capacitance probes have additional electronics to offset material buildup on the probe.

Special probes are available that produce a signal proportional to flow in open channel flumes and weirs. These probes are characterized to match the head/flow relationship of an open channel primary element. Characterization is accomplished either of two ways: one, by electronic calculation, or two, by variation of probe insulation. Thickness in a manner that produces a direct, linear relationship between capacitance and flow. Consult with manufacturers on special capacitance probes for direct flow measurement.

Capacitance probes can also be used to measure the level of dry material. However, this discussion deals with only liquid level measurement applications. Application guidelines for uncompensated and compensated capacitance probes are shown in Tables 5.2 and 5.3, respectively.

Principle of Operation

A capacitor consists of two electrically conductive plates separated by a nonconductive material. A probe is usually constructed to form one plate of a capacitor. The other plate is the tank wall or the measured solution. Between the probe and the tank wall is an air space above the liquid surface and water below. As the water level rises, the effective capacitance of the system

Table 5.2. Uncompensated Capacitance Probes Application Guidelines

Recommended	Not Recommended
Potable water	Primary effluent
Noncoating chemicals	Secondary effluent
Tertiary effluent	Sludges
Raw water	Polymer solutions
Finished water	Lime slurry

Table 5.3.   Compensated Capacitance Probes Application Guidelines

Recommended	Not Recommended
Most aqueous solutions	Liquids where heavy grease buildup
Raw sewage	could occur
Primary effluent	
Secondary effluent	

increases. This capacitance is linearly proportional to level and is measured by a bridge circuit powered by a high frequency, 0.5–1.5 MHz, oscillator. High frequency can reduce errors due to shorting of the capacitor by conductive coatings. Sometimes capacitance probes are referred to as radio frequency (RF) probes because of this measurement technique.

Water is a good conductor. For this reason, the probe must be insulated. The insulation's exterior surface effectively becomes a third plate which complicates the theory of operation. Figure 5.3 shows one way of illustrating the system in electrical terms.

The situation becomes more complex if a conductive coating of process solids accumulates on the probe. Through a combination of capacitance and conductance effects in the coating, the probe fails to respond to changes in level below the top of the coating. The system's effective capacity remains constant below this point. Thus, for most applications in wastewater treatment some method of compensating for coatings is essential.

One method is based on assuming that the coating's capacitive reactance is equal to its resistive reactance. The coating's resistance is further assumed to be the greatest resistance in an aqueous system. This resistance is measured and subtracted from the effective capacitance. The result is proportional to liquid level, although inaccuracies are introduced depending on how well the system matches the assumptions made.

The probe is usually a cylindrical rod or cable inserted perpendicular to the water surface, as illustrated in Figure 5.4.

Probes are available for use in open-channel head-loss type flow meters. These probes are flat and shaped to provide a signal proportional to flow. In either style probe, accuracy decreases near the bottom because the submerged portion becomes less and less like the ideal capacitor with plates of infinite length.

Accuracy and Repeatability

Accuracy   ±1% of full scale.
            Where the probe becomes coated, accuracy will degrade to approximately ±5%

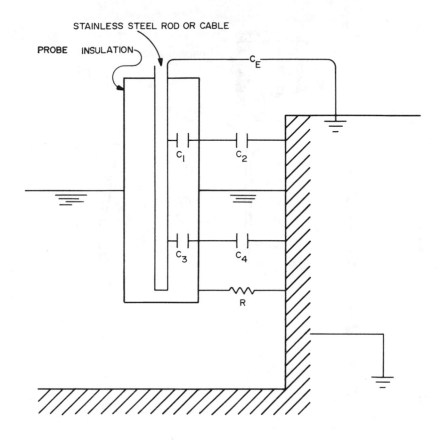

Figure 5.3. Probe/tank capacitive relationship.

Manufacturers' Options

1. Probe type:
    a. rods of any length up to 6 m (20 ft)
    b. cables of any length up to about 50 m (150 ft) with weight or anchor to keep probe in place
    c. flat probes for open channel flow meters
    d. proximity plate for noncontact with process fluids
    e. integral transmitter/probe. Transmitter mounted within probe head
2. Transmitter enclosure:
    a. NEMA 3 – weatherproof
    b. NEMA 4 – watertight
    c. NEMA 7 – explosion-proof

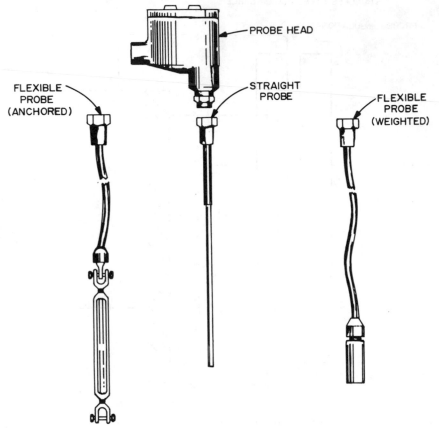

Figure 5.4.   Capacitance level sensor.

3. Indicating meter
4. Output:
   a. 4–20 mAdc
   b. 10–50 mAdc
5. Probe materials:
   a. 304 stainless steel
   b. 316 stainless steel
   c. Teflon insulator
   d. polyvinylidene fluoride insulator
   e. polyvinylchoride insulator
6. Grounding rods
7. Concentric probe. Required for nonconductive liquids and for some

installations to provide grounding or shielding from process liquid turbulence

8. Radio frequency interference protection

Installation

In an open tank, install a depth (staff) gauge in the tank at a location visible from the transmitter so periodic checks can be made on the calibration. Zero on this gauge must correspond to the probe's zero reference.

Probe

- Isolate from vibration and possible physical damage.
- Do not mount in the direct stream of process flow. If necessary, install baffles or stilling well.
- Mount vertically.
- Mount at least 15 cm (6 in.) from tank wall to lessen chances of material buildup.
- Mount so that probe can be removed readily for cleaning, inspection, or calibration.

Transmitter

Install the transmitter within 75 m (250 ft) of the probe. The cable must be furnished by the capacitance probe manufacturer and should not be shortened in the field without consulting the manufacturer. Mount the transmitter close enough to the probe to see indicator change as level changes, to allow one person to calibrate the system. If the transmitter is mounted farther than this, provide a junction box near the probe. Allow enough cable between the junction box and probe for removing the probe. Provide storage space for excess cable. Cable from the junction box to the transmitter should be in rigid conduit and from the junction box to the probe in flexible conduit. Figure 5.5 illustrates capacitance probe and transmitter installation.

Designer Checklist

If you can answer "yes" to the following questions when designing or reviewing capacitance level measurement applications, the application should be correct.

- Is the process stream free from heavy grease? If not, capacitance probes are not recommended.
- Is coating of the probe likely? If yes, then coating compensation is essential.

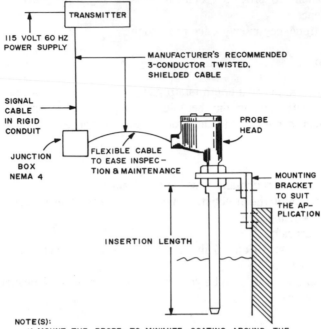

Figure 5.5.   Capacitance probe and transmitter installation.

- Is the tank or tank wall grounded? If not, then provide grounding.
- Is the probe mounted securely, without providing potential sites for solids buildup?
- Can the probe be removed easily for inspection, cleaning, and calibration?
- Is the meter installation designed so that it can be calibrated by one person?
- Is the transmitter protected from the weather?
- Has a depth gauge been installed for quick calibration checks?

Maintenance and Calibration

Task	Frequency
1. Clean probe	Check daily or weekly until history is established

2. Calibrate probe   Once a month to once every
                                       two months

Deficiencies

The following problems are often encountered in capacitance probe applications.

- Coating causes meter to measure level inaccurately. Clean probe more frequently.
- Tank not grounded signal noise and calibration drift.
- Probe too long to be easily removed for maintenance.

FLOATS

Application

Float type level indicators are often used in treatment plants if remote readout is not needed. Floats can be connected to transmitters for remote monitoring. However, this arrangement is seldom used. Some other type of level meter such as bubbler, capacitance, or sonic is used instead.

Float type level switches are generally used for alarms and equipment on/off control.

Application guidelines for float level indicators are listed in Table 5.4.

Principle of Operation

Level Indicator

A float level indicator consists of a float, an attached rod with pointer, float guide, and indicator scale. These components are shown in Figure 5.6. As the float rides up or down on the liquid surface, the pointer indicates the level.

Another type of float level indicator is shown in Figure 5.7. In this case, float movement is indicated by the counterweight position.

Table 5.4. Float Level Indicators Application Guideline

Recommended	Not Recommended
Finished water	Liquids where grease or solids
Tertiary effluent	buildup could occur
Raw water	Polymer solutions
Potable water tanks	Turbulent processes
Fuel tanks	

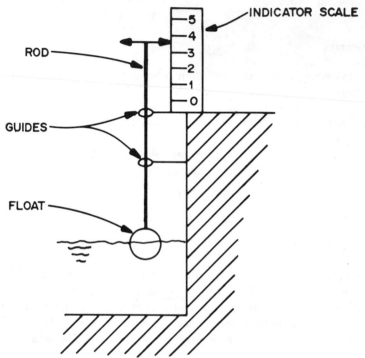

Figure 5.6.   Simple float level indicator.

Level Switch

Float switches depend on the liquid's buoyant force to activate the switch. In one switch position the float is buoyed up by the liquid; in the other position the float hangs down in the absence of liquid.

A wide variety of float devices exist which translate the float position into electrical signals. The *Instrument Engineers' Handbook* has a summary of them (see Bibliography). Of these devices, the majority of water and wastewater applications use the tilt switch type shown in Figure 5.8. Each switch is a buoyant bag with a mercury switch inside. When the bag is tilted from one position to the other, the mercury switch opens or closes an electrical circuit. This circuit activates a relay to provide the contacts necessary for local controls and remote monitoring.

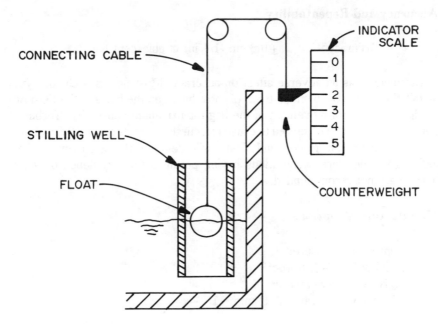

Figure 5.7.   Counterweighted float level indicator.

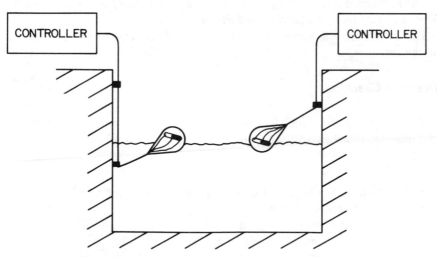

Figure 5.8.   Float switches.

Accuracy and Repeatability

Accuracy ± 0.6 cm (1/4 in) in quiescent liquids

Turbulence has an adverse affect on accuracy. Most float switch (tilt type) installations leave 8–16 cm (3–6 in.) of cable between the bag and the tie-down to allow freedom of movement of the bag for maximum sensitivity. Turbulent conditions have been observed to cause inaccuracies greater than 10 cm (4 in.).

Solids buildup on guides and floats will degrade the accuracy of float indicators. Therefore, applications having potential for turbulence or solids buildup are not recommended.

Manufacturers' Options

1. Controller enclosure:
   a. NEMA 12 – dustproof
   b. NEMA 4 – weatherproof
2. DPDT contacts rated at 5 amp, 115 Vac

Installation

A recommended installation method for float switches is shown in Figure 5.9. If turbulence is expected, also install a stilling well.

The installation figure shows the floats permanently fixed to a specific level. This is true of most float switch installations.

The switch setting cannot be easily changed, so carefully design initial placement.

Designer Checklist

If you can answer "yes" to the following questions when designing or reviewing applications of float level instruments, the application should be correct.

- Will the float be buoyant at process pressure? Temperature?
- Is the float protected from turbulence?
- Is the output matched to its intended uses?
- Are switch settings accurately known?
- Are switches sheltered from strong air currents?
- Is the process stream free from heavy grease?
- Is coating of the float likely? If yes, floats are not recommended.

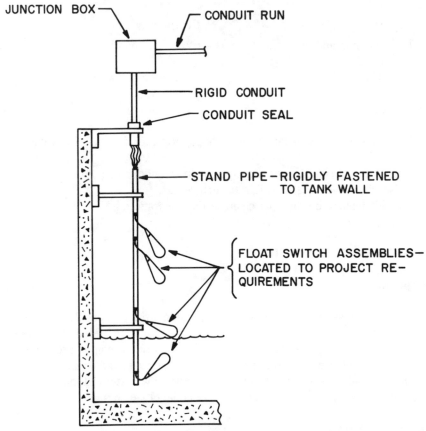

Figure 5.9.   Float switch mounting installation.

Maintenance and Calibration

Level Indicators

Task	Frequency
1. Calibration	Depends on mechanical or electrical linkage to indicator
2. Stilling well cleaning	Every one or two weeks

Level Switches

Task	Frequency
1. Inspection and operational check	Every six months
2. Stilling well cleaning	Every one or two weeks

Deficiencies

The following problems are commonly reported for float level devices.

- Stilling well cleaning requires too much attention.
- Switches give false trips due to turbulence.
- Solids buildup on the float changes the calibration.

SONIC AND ULTRASONIC

Application

Sonic and ultrasonic level sensors do not contact the process fluid. They can be used in most water and wastewater applications, provided that process vapors do not cause problems. The two most common vapor problems are corrosion, which can be lessened by choice of materials, and condensate or ice buildup on cold sensors, which can be prevented by heaters. Table 5.5 shows application guidelines for sonic and ultrasonic level sensors.

Operating Principle

The sensor periodically generates a pulse of sonic or ultrasonic waves that bounce off the liquids' surface and echo back. The echo is detected by a resonant metal disk. Based on the speed of sound or ultrasound, the time between sending and receiving is measured and converted into distance which is then converted to level (see Figure 5.10).

Sonic and ultrasonic wave velocity depends on air temperature, pressure, and humidity. Where changing conditions are expected, automatic compensation can

Table 5.5.   Sonic and Ultrasonic Level Sensor Application Guidelines

Recommended	Not Recommended
Open channel flow Wet wells, reservoirs	Liquids with surface foam

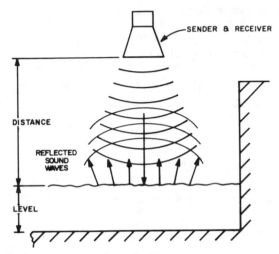

Figure 5.10.   Sonic sensor operating principle.

be provided. Typically, only temperature compensation is used. Temperature errors are about 0.2% per °C (0.17% per °F).

Sensors are available with frequencies from about 9 kHz sonic to about 50 kHz (ultrasonic). Also, the generator can have different shapes such as wide-angle cone, narrow-angle cone or parabolic. Selection of frequency and sensor shape are both based on the amount of attenuation expected. For example, one manufacturer provides:

1. wide angle cones for up to 3 m (10 ft)
2. narrow angle cones for up to 10 m (30 ft)
3. parabolic reflectors for up to 25 m (80 ft)

Signal attenuation can be caused by absorption into the air, reflection away from the receiver's sensing area, and absorption by foam on the liquid surface. The cone shapes listed above are selected to reduce attenuation by reflection. Distance and wave frequency affect attenuation by absorption. As distance from the sensor to the liquid level increases, signal strength decreases in proportion to the distance squared. Thus, if signal strength is 100% at distance "d" when a tank is full, the signal strength will drop to 25% at a distance of "2d".

Sonic waves attenuate less than ultrasonic. For example, foam on liquid surfaces may completely absorb ultrasonic waves.

Accuracy and Repeatability

Accuracy	$\pm 1\%$ of span
Repeatability	$\pm 0.1\%$ of span

Air conditions, liquid turbulence, foam, and interfering echoes from obstructions can reduce both accuracy and repeatability of the sensor.

Manufacturers' Options

1. Sensor wave frequency:
   a. sonic
   b. ultrasonic
2. Sender shape:
   a. wide cone
   b. narrow cone
   c. paraboloid
3. Sensor temperature compensation
4. Sensor thermostatically controlled heater
5. Sensor air purge
6. Sensor range selection
7. Indicator on transmitter
8. Transmitter output:
   a. 4–20 mAdc
   b. 0–20 mAdc
9. Transmitter enclosure:
   a. NEMA 1
   b. NEMA 4
   c. NEMA 12

Installation

Determine the range of the meter from expected conditions in the tank or channel (see Figure 5.11). The mounting location of the sensor is then calculated from restrictions established by the manufacturer. Generally, the sensor must be above the maximum level by at least some minimum distance, usually about 50–70 cm (18–24 in.).

The sensor must be mounted far enough from tank walls to prevent false echoes. This distance depends on sender shape. Calculations for correct sensor location will differ for each manufacturer.

Where overhead clearance will not allow the minimum distance to be met, the sensor can be mounted to project a horizontal beam at a stainless steel plate mounted at a 45° angle to the beam.

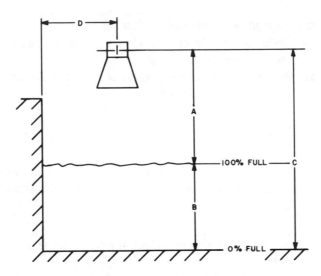

A   = Distance from generator/receiver to 100% full level
B   = Measured range, distance from 0 to 100% full
C   = A + B
D   = Distance from tank wall
Manufacturer's set limits on dimensions of A, C, D and the ration A/C

Figure 5.11.   Installation dimensions.

Stilling Well

A stilling well is used with sonic/ultrasonic sensors to dampen out liquid level turbulence, reduce foam, increase signal strength (essentially producing a cylindrical-shaped sensor), eliminate noise from stray echoes or to lessen condensate problems. When used, the stilling well should be cut from a single length of PVC pipe 15–20 cm (6–8 in.) in diameter. The bottom end should be cut at a 45° angle.

Drill air relief holes near the top where the sensor is attached.

The stilling well must be kept clean. Accumulated solids can cause echoes that the transmitter will read as liquid level. Therefore, provide for either manual or automatic washdown of the well interior wall.

Transmitter

The transmitter location depends on the intended method of calibration. The transmitter may be remotely located as much as 200 m (700 ft) away provided that:

- The sensor is equipped with a calibration bar, or
- The tank can be isolated from feed and exit streams and the level manually raised (or lowered), or
- The liquid level in the tank can be manually observed at the transmitter location by an independent method.

Designer Checklist

If you can answer "yes" to the following questions when reviewing or designing a sonic ultrasonic level meter application, the application should be correct.

- Is temperature compensation provided? If not, is the degraded accuracy acceptable?
- Is condensation unlikely to occur on the sender/receiver? If it is likely, then is an air purge and/or heater provided?
- Can a stilling well be avoided?
- Can the sensor be mounted so that the full range of expected levels are within the manufacturer's specifications for minimum and maximum distances?

Maintenance and Calibration

Task	Frequency
1. Calibration	Every two months
2. Check temperature compensation	Every two months
3. Stilling well cleaning	Depends on process stream

Deficiencies

The following problems have been reported for sonic and ultrasonic level meter installations.

- Stilling well causes interfering echoes from pipe joints or deposited solids or grease. Maintaining a clean stilling well is difficult.
- Meter does not read on cold days because receiver is covered with ice.

BIBLIOGRAPHY

Bubblers

1. Liptak, B. G., and K. Venczel. *Instrument Engineers Handbook*, rev. ed. (Radnor, PA: Chilton Book Company, 1982).
2. Manual on Installation of Refinery Instruments and Control Systems, Part I, Section 6–Level, 3rd ed., (Washington, DC: American Petroleum Institute, 1974).

Capacitance Probes

1. Liptak, B. G., and K. Venczel. *Instrument Engineers Handbook*, rev. ed. (Radnor, PA: Chilton Book Company, 1982).
2. Schuler, E., *A Practical Guide to RF Level Controls* (Horsham, PA: Drexelbrook Engineering Co., 1981).

Floats

1. Liptak, B. G., and Venczel, K. *Instrument Engineers Handbook*, rev. ed. (Radnor, PA: Chilton Book Company, 1982).
2. Instrumentation in Wastewater Treatment Plants; Manual of Practice No. 21 (Washington, DC: Water Pollution Control Federation, 1978).

Chapter 6

Pressure Measurement

PRESSURE CELLS

Application

Pressure meters are applied to enclosed process lines such as compressed air distribution systems, pump discharges, and tanks. With the aid of isolation diaphragms, pressure transmitters can be successfully applied to any water or wastewater treatment process. Pressure measurement application guidelines are shown in Table 6.1.

Principle of Operation

Mechanical Pressure Elements

The three most common elements used to measure pressure are Bourdon tubes, bellows, and diaphragms. In each case process pressure causes the element to move in proportion to the pressure applied. This motion is amplified by a mechanical linkage connected to a pointer and dial or by electronics to a voltage or current signal. Schematic diagrams of each type are shown in Figure 6.1.

Table 6.1. Pressure Measurement Application Guideline

Recommended	Recommended with Isolation Diaphragm
Air	Chlorine
Oxygen	Wastewater with solids
Digester gas	Sludge
Water (raw and finished)	
Secondary effluent	

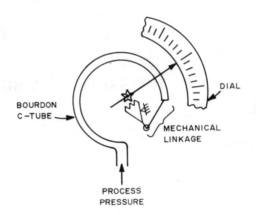

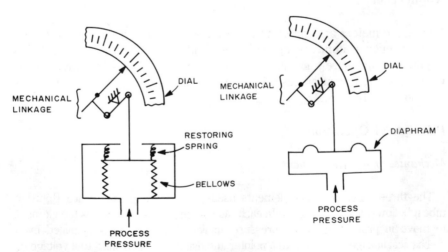

Figure 6.1. Mechanical pressure elements.

Bourdon tube.   A Bourdon tube is a curved tube sealed at the tip. As process pressure increases inside the tube, the tube will straighten, causing the tip to deflect. The deflection is transferred to a dial indicator by mechanical linkage. Besides a C-shaped tube, Bourdon tubes are available in spiral, twisted, and helical forms, round, oval, or rectangular in cross section.

Bellows.   Bellows elements are deeply corrugated metal cylinders closed at one end. Process pressure applied to the bellows causes it to expand. Bellows expansion is transferred to a dial indicator by mechanical linkage. Bellows are also configured to contract on increasing pressure. In some cases restoring springs are added to increase operating range or to reduce element wear.

Diaphragms.   Diaphragms are metal disks, either flat or concentrically corrugated. Process pressure applied to one side causes the diaphragm to deflect outward. Diaphragm deflection is transferred to a dial indicator by mechanical linkage. Corrugated diaphragms are capable of greater deflection and are more linear than flat diaphragms.

Electro Mechanical Elements

Electric signals proportional to pressure are obtained by mechanically connecting an electrical component such as a capacitor, strain gauge, or inductor to a diaphragm. Deflection of the diaphragm will change the associated electrical property; e.g., distance between plates in a capacitor, piezoelectric response, and loop reluctance.

In some elements a restoring force is applied to the diaphragm to keep it undeflected. This eliminates nonlinearity due to diaphragm deflection. The restoring force is measured electrically and converted to pressure. These elements are called force-balance transducers.

Gauge, Differential, and Absolute Pressures

Gauge, differential, and absolute pressures may be measured by any pressure element, depending on the reference pressure. Gauge pressure is measured using atmospheric pressure as reference. In this case, Bourdon tubes, bellows and diaphragms are constructed to have access to the atmosphere on the side opposite the process connection.

Vacuum Measurement

Vacuum pressures for slight vacuums, to –90 kPa, can be measured with elements similar to those discussed above. At low vacuums these elements become inadequate because of gravitational interference in force measurement near zero atmosphere. Usually, low vacuums are measured indirectly from some other property of gas such as thermal conductivity, viscosity, or the behavior of gas during ionization or electrical discharge. Low vacuums are rarely monitored in water or wastewater applications. Therefore, their measurement is not discussed further.

Accuracy and Repeatability

Quartz Bourdon Tubes

> Accuracy   ±0.01% of span
> as accurate as a quality manometer

On-line Pressure Transmitters

> Accuracy   ±0.5% of span

Pressure Gauges

> Accuracy   ±1 to ±2% of span

Typically, repeatability of measurement is about one-fifth of the stated accuracy.

Manufacturers' Options

1. Ranges — almost any range is available
2. Materials (wetted):
   a. brass
   b. bronze (or phosphor bronze)
   c. beryllium copper
   d. stainless steel
   e. Monel
   f. Hastelloy C
3. Transmitter output signals:
   a. 4–20 mAdc
   b. 10–50 mAdc
   c. 1–5 Vdc
4. Power:
   a. two-wire transmitters are the most common configuration; they require external 12 Vdc or 24 Vdc power supplies
   b. four-wire transmitters which require line power (115 Vac, 60 Hz)
5. Transmitter enclosure:
   a. NEMA 4

b. explosion-proof

6. Isolation diaphragms of the same wetted materials listed for the elements are available from some manufacturers. Diaphragm, transmitter, and connection line can be furnished as a complete assembly.

Installation

- Install the transmitter in an environment that meets the specifications listed by the specific manufacturer. This is usually –20 to 65°C (0 to 150°F) and 0 to 95% relative humidity. Zero and span will shift with ambient temperature, so avoid temperature extremes or calibrate at conditions equal to the installed environment.
- Install the transmitter as close as possible to the process measurement site. This will reduce response time which can be important in flow control or level control applications. The installation must allow good maintenance access. In some cases it will not be practicable to install the transmitter to meet both nearness and maintenance criteria. In these situations, control requirements must be given first priority.
- Connect meter runs to liquid process lines horizontally. This will minimize the amount of solids and gas entering the connection. Entrapped gas will decrease response time and solids may plug the meter connection. Slope meter runs 8 cm per meter (1 in. per ft) of run, so that gas bubbles bleed back into the process line.
- Connect meter runs to gaseous process lines at the top of pipes or tanks to minimize the amount of solids and moisture entering the connection. Slope meter runs at least 8 cm per meter (1 in. per ft) of run so that condensation will drain into the process line. Any low spots in the meter run will require a condensate collection pot. Heat trace meter runs on condensable gases. Entrapped liquids may affect meter accuracy and may cause accelerated corrosion.
- For applications where the measured liquid contains solids, flushing provisions or diaphragm isolation may be needed. Diaphragm connections to the process should be a minimum of 2.5 cm (1 in.) for sludge lines and 1.3 cm (0.5 in.) diameter for other wastewater lines.
- Install an isolation valve at the process measurement connection on all meter runs. If this valve is not readily accessible for maintenance, install another isolation valve at the transmitter.
- Materials recommended for harsh environments are:
  a. chlorine–Hastelloy-C
  b. digester gas–316 stainless steel

Designer Checklist

If you can answer "yes" to the following questions when designing or reviewing pressure meter applications, the application should be correct.

- Is the meter situated for adequate response time and good maintenance access?
- Are meter runs installed to keep out interfering substances?
- Can the meter be calibrated in place?
- Is the meter in a suitable environment?

Maintenance and Calibration

Task	Frequency
1. Calibration	Every three to six months Meters used in critical control applications may need more frequent calibration

Deficiencies

The following problems are commonly reported for pressure transmitters:

- Meter installed in an inaccessible location
- Meter runs incorrectly installed
- Diaphragms or flushing not provided on sludge lines

DIFFERENTIAL PRESSURE

Application

Differential pressure transmitters, DP cells, are used with primary elements to measure flows, gauge pressure, and liquid level. With the aid of isolation diaphragms or purge systems, DP cells can be applied successfully to any water or wastewater treatment process. Application guidelines for differential pressure are in Table 6.2.

Table 6.2. Pressure Measurement Application Guideline

Recommended	Recommended with Isolation Diaphragm
Air	Chlorine
Oxygen and ozone	Wastewater with solids
Digester gas	Sludge
Water (raw and filtered)	
Secondary effluent	

Principle of Operation

Mechanical Pressure Elements

The three most common elements used to indicate pressure are: Bourdon tubes, bellows, and diaphragms. In each case, the element moves in proportion to differential pressure. This motion is amplified by mechanical linkage to a pointer and dial. Schematic diagrams of each type are shown in Figure 6.2.

Bourdon tube. A Bourdon tube is a curved tube sealed at the tip. As process pressure increases inside the tube, the tube straightens, causing the tip to deflect. The deflection is indicated on a dial by mechanical linkage. Besides a C-shaped tube, Bourdon tubes are available in spiral, twisted, and helical forms, round, oval, or rectangular in cross section.

Bellows. Bellows elements are deeply corrugated metal cylinders closed at one end. Process pressure applied to the high side of the bellows causes it to expand. Bellows expansion is converted to pointer and dial indication. Bellows are also configured to contract on increasing pressure. In some cases restoring springs are added to increase operating range or reduce element wear.

Diaphragms. Diaphragms are metal disks, either flat or concentrically corrugated. High and low process pressures are applied to opposite sides of the diaphragm. This causes the diaphragm to deflect. A mechanical linkage connects the diaphragm to a pointer for dial indication. Corrugated diaphragms allow larger deflection and better linearity than flat diaphragms.

Electro-Mechanical Elements

Electric signals proportional to differential pressure are obtained by mechanically connecting an electrical component such as a capacitor, strain gauge, or inductor to a diaphragm. Deflection of the diaphragm will change the associated electrical property; e.g., distance between plates in a capacitor, piezoelectric response, and loop reluctance.

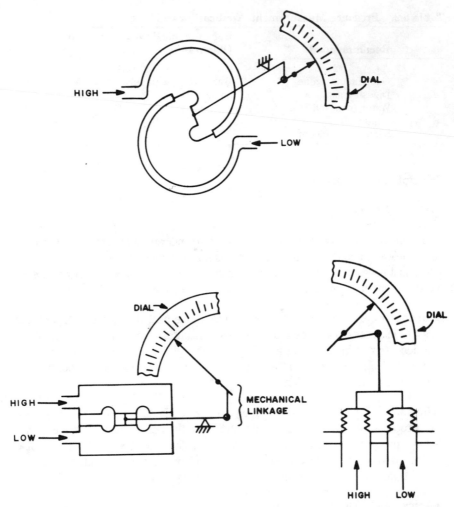

Figure 6.2. Mechanical differential pressure elements.

In some elements a restoring force is applied to the diaphragm to keep it undeflected. This eliminates nonlinearity due to diaphragm deflection. The restoring force is measured electrically and converted to differential pressure. These elements are called force-balance transducers.

Absolute Pressure Elements

Absolute pressure elements are differential pressure elements with the low pressure side evacuated to −101.3 kPa and sealed from atmosphere.

Accuracy and Repeatability

Quartz Bourdon Tubes

Accuracy $\pm 0.01\%$ of span
equal to the accuracy of a good
quality manometer

On-line Differential Pressure Transmitter

Accuracy $\pm 0.5\%$ of span

Differential Pressure Gauges

Accuracy ± 0.5 to $\pm 2\%$ of span

Repeatability of measurement for differential pressure gauges and transmitters is commonly about one-fifth of the rated accuracy.

Manufacturers' Options

1. Ranges—almost any range is available
2. Materials (wetted):
   a. brass
   b. bronze (or phosphor bronze)
   c. beryllium copper
   d. stainless steel
   e. Monel
   f. Hastelloy C
3. Transmitter output signals:
   a. 4–20 mAdc
   b. 10–50 mAdc
   c. 1–5 Vdc
4. Power:
   a. two-wire transmitters are the most common. They require a sepa-rated 12 Vdc or 24 Vdc power supply.
   b. four-wire transmitters which require line power (115 Vac, 60 Hz)
5. Transmitter enclosure:
   a. NEMA 4
   b. Explosion-proof

    c. both available with insulated jackets or boxes

6. Isolation diaphragms of the same wetted materials are available from some manufacturers. Diaphragms, transmitter, and connection line can be furnished as a complete assembly.
7. Square root extractor
8. Scales:
    a. linear
    b. square root

Installation

- Install the transmitter in an environment recommended by the manufacturer. This is usually –20 to 65°C (0 to 150°F) and 0 to 95% relative humidity. Zero and span will shift with changes in temperature, so avoid temperature extremes.
- Install the transmitter as close as possible to the process measurement site to reduce response time which can be important in flow control or level control applications. The installation must allow easy access for maintenance. In some cases it will not be practical to install the transmitter to meet both nearness and maintenance criteria. In these situations, control requirements must be given first priority.
- For solids-bearing liquid process lines, connect meter runs horizontally. Do not connect meter runs to the upper quadrant of the pipe. This will minimize the amount of solids and gas entering the connection. Entrapped gas will decrease response time, and solids may plug the meter connection. Slope meter runs 8 cm per meter (1 in. per ft) of run so that gas bubbles bleed back into the process line.
- Connect meter runs to gaseous process lines at the top of pipes or tanks to minimize the amount of solids and moisture entering the connection. Slope meter runs at least 8 cm per meter (1 in. per ft) of run, so that condensation will drain into the process line. Low spots in meter runs should be avoided; if they cannot, add drain pots to these low spots. Heat trace meter runs on condensable gases. Entrapped liquids will affect meter accuracy and may cause accelerated corrosion.
- Special precautions must be taken in steam applications to prevent overheating of the manifold and transmitter. Side mount pressure taps to allow steam into the tap while still allowing drainage of excess condensate back into the process pipe. A condensate pot should be installed on each meter lead. To avoid overheating, blowdown valves should not be incorporated in the manifold.

    In steam applications, a water seal is required between the condensate pot and the manifold. This prevents steam from reaching the manifold and transmitter and prevents uncontrolled buildup of condensate in the meter leads. The condensate pots must be identical in size, the same height above the transmitter, and self-draining to the process pipe.

- For applications involving solids-bearing liquids, flushing provisions or diaphragm isolation may be needed. Diaphragm connections to the process should be a minimum of 2.5 cm (1 in.) for sludge lines.

- Install an isolation valve on all meter runs at the process measurement site (pressure tap) and, except for very short tap lines, at the transmitter.

- Materials recommended for harsh environments are:
  a. chlorine – Hastelloy
  b. digester gas – 316 stainless steel

- Differential pressure transmitters are usually installed with valves that enhance some combination of instrument calibration, blowdown of accumulated material in the meter piping, and isolation of the transmitter. An economic way to provide the desired functions is to use factory-made valve manifolds. Some common manifolds are shown in Figure 6.3. Available manifold options are:
  a. number and configuration of valves:
     1) 3-valve: isolation and equalization
     2) 5-valve: isolation, equalization, and calibration
     3) 5-valve: isolation, equalization, and blowdown
     4) 7-valve: isolation, equalization, calibration, and blowdown
  b. process connections:
     1) pipe: 1/2" NPT female and 3/8" NPT female
     2) tube: 3/8" and 1/2"
  c. transmitter connections:
     1) pipe: 1/2" NPT female
     2) direct-flanged
     3) tube: 3/8" and 1/2"
  d. materials of construction:
     same selection as for transmitter.
  e. remote zeroing: motor operated 3-valve manifold
     A check of meter zero is sufficient and a check of the span is not required at a frequent enough interval to justify the expense of a more complex manifold. Span calibration test frequencies of three to six months for most applications confirms this practice. Where DP cells are part of a flow measurement system used as a standard, or for billing purposes, the frequency of calibration and testing may be much more often. In such cases it is recommended that 5-valve manifolds be installed to reduce calibration setup time.

- A common use of DP cells is with primary flow elements to measure pressure-drop for flow calculation. Special installation practices for flow applications are presented with the primary device; see orifice meters, venturi meters.

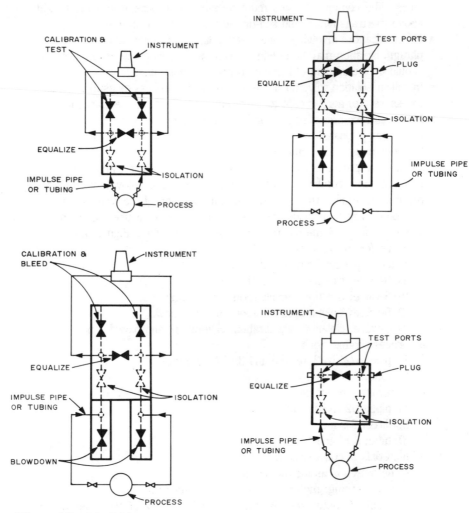

Figure 6.3. Manifolds.

- Another common usage of DP cells is to measure liquid levels. Two general methods are used: hydrostatic head and bubbler (dip tube). Bubbler installations are discussed in Chapter 5. Figure 6.4 shows typical hydrostatic level installations for an enclosed tank. The tank is covered, so the DP cell must use the pressure of the vapor phase as a reference. The DP cell is at the same level as the bottom pressure tap, and the connection to the top tap is vapor filled or "dry." Thus, the difference in pressure is proportional to liquid level. If the tank was open, a plumbing connection to

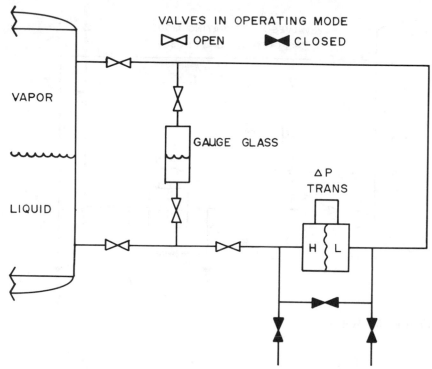

VALVES IN OPERATING MODE

⋈ OPEN   ◀▶ CLOSED

VAPOR

GAUGE GLASS

ΔP
TRANS

LIQUID

H   L

Figure 6.4. Dry leg.

the reference side of the DP cell would not be necessary. The DP cell would just need to be at the same ambient pressure, i.e., in the same room with the tank or both the DP cell and the tank outdoors.

A "wet leg" configuration for level measurement is shown in Figure 6.5. The reference side of the DP cell is filled with a liquid. It does not have to be the same as the tank liquid. If the liquid is not the same, the difference in specific gravity of the two liquids must be used to correct the meter calibration. The liquid in the wet leg prevents unwanted accumulation of condensate at the reference side. In this configuration the reference (wet leg) is connected to the low-pressure side of the DP cell, just as in a dry leg setup. As shown, the meter would read: 100% at '0' liquid level and '0'% at 100% liquid level. This is corrected during calibration by suppressing zero to provide a correct tank level indication.

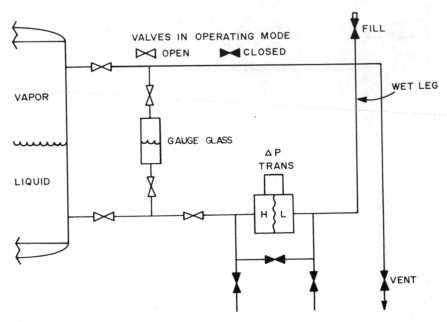

Figure 6.5. Wet leg.

Designer Checklist

If you can answer "yes" to the following questions when designing or reviewing differential pressure meter applications, the application should be correct.

- Is the meter situated for adequate response time and good maintenance access?
- Are meter runs installed to keep out interfering substances?
- Can the meter be calibrated in place?
- Is the meter in a suitable environment?

Maintenance and Calibration

Task	Frequency
1. Calibration	Every three to six months Meters used in critical control applications may need more frequent calibration

Deficiencies

The following problems are commonly reported for pressure transmitters:

- Meter installed in an inaccessible location.
- Meter runs incorrectly installed.
- Diaphragms or flushing not provided on sludge lines.

BIBLIOGRAPHY

Pressure Cells

1. Gillum, D. R., *Industrial Pressure Measurement* (Research Triangle Park, NC: ISA Publications, 1982).
2. Hewson, J. E., *Process Instrumentation Manifolds* (Research Triangle Park, NC: ISA Publications, 1981).
3. "Measurement & Control News, Pressure/Force Handbook and Buyers Guide, 1983." (Pittsburgh, PA: Measurements & Data Corporation, 1982).

Differential Pressure

1. Hewson, J.E., *Process Instrumentation Manifolds* (Research Triangle Park, NC: ISA Publications, 1981).
2. "Manual on Installation of Refinery Instruments and Control Systems, Part I — Process Instrumentation and Control, Section 1 — Flow, and Section 2 — Level," Washington, DC: American Petroleum Institute, 1974).
3. "Measurement & Control News, Pressure/Force Handbook and Buyers Guide, 1983," (Pittsburgh, PA: Measurements & Data Corporation, 1982).

Chapter 7

Quality Assurance Testing

OBJECTIVES

Instruments that are properly selected, installed, and maintained can be very effective in improving overall plant operation. Improvements can be in the treatment process such as more efficient chemical and energy use, increased security, or more consistent effluent or finished water quality. Instruments can also provide more timely and accurate process information to operators and managers.

The objectives of quality assurance testing are as follows:

1. To obtain instruments which are accurate, reliable, and stable and which do not require excessive maintenance.
2. To verify the instrument meets specified acceptance criteria for the particular application.
3. To determine the long term maintenance requirements.
4. To provide a calibration reference point.

Even if an instrument is factory calibrated prior to delivery, perform in-place acceptance testing. Monitor instrument performance as an ongoing practice.

Note that the word instrument means instrument system. For example, a flume consists of the primary element, the flume; a secondary element, the level sensor; and the transmitter which converts level to flow. Acceptance testing compares the flow reading of the transmitter against a reference flow, not just the transmitter output based on simulated level input or the level sensor/transmitter combination.

ACCEPTANCE TESTING

Instruments should be designed for realistic operating conditions and they should be evaluated under the same conditions. However, it is not always practical to evaluate performance under all possible operating conditions.

The Scientific Apparatus Makers Association, SAMA, has published generic test methods for the testing and evaluation of process measurement and control instrumentation. These test methods are used mainly be instrument manufacturers under laboratory conditions. The methods measure the effect of ambient temperature, atmospheric pressure, relative humidity, line voltage, and frequency as well as accuracy, repeatability, linearity, and related parameters.

The Instrument Testing Service, Inc., ITS, has established test protocols for various waste water plant instruments. The tests, which follow the SAMA test methods, include both bench test procedures and field test procedures. Bench tests are up to 80 hours to determine drift. Field testing is conducted for 60 consecutive days. The ITS has published results for chlorine residual analyzers. It is currently testing dissolved oxygen analyzers and preparing a suspended solids analyzer test protocol.

Acceptance Test Procedure

The following test procedure is based on SAMA Standard PMC 31.1–1980. Under actual operating conditions, simplified evaluation procedures may have to be used for calibration and acceptance testing.

1. Prior to calibration, be sure that the instrument is properly installed and that it is given a preliminary checkout according to the manufacturer's instructions.
2. Zero the instrument and correct any offset according to manufacturer's calibration instructions.
3. Simulate the signal produced by the primary element and input to the converter/transmitter to set the zero and span.

4. Set up the process so that the variable to be measured is near the midrange of the instrument. Allow enough time for the process to become steady, then measure the reference variable (V_R). Use one of the methods described later in this chapter.
5. During the time V_R is being measured, record readings from the instrument. Calculate the average of the recorded readings (V_A)
6. Calculate the percent error using the following equation:

$$\%\text{Error} = [(V_A - V_R)/V_R] \times 100$$

7. Record V_R, % Error as shown in Table 7.1.
8. Repeat steps 4 through 7 for a minimum of eight more readings within the design range of the instrument. Starting from the midrange reading, move upscale to 75% and 90% of maximum. Move downscale to 75%, 50%, 25%, and 10% and then back up to 25% and 50%. Repeat to determine repeatablilty.
9. Calculate the average of the upscale and downscale tests and the overall average.

The accuracy is the largest upscale and downscale reading. In the example table, the measured accuracy is +0.23/-0.32. The repeatability is the largest difference between any upscale or downscale test. In the example table it occurs on the downscale test at 50% and is equal to 0.03.

If the measured accuracy or repeatability is greater than the specification requirements for that type of instrument, the performance of the instrument may be unsatisfactory and the following options should be considered:

1. If the measured accuracy is constant, a span adjustment at the instrument transmitter is required. Follow the manufacturer's instructions. Repeat the testing to verify correct span.
2. If the instrument cannot be adjusted to meet specifications, repair or reject under acceptance specifications.
3. If poor instrument performance is due to nonideal installation conditions (e.g., inadequate approach piping for mag meters) and the data

Table 7.1. Example Table of Test Results

V_R	up actual	down actual	up actual	down actual	up actual	up avg	down avg	avg error
				%Error				
10		+0.14		0.15			+0.15	+0.15
25		+0.23	+0.08	+0.21	+0.10	+0.09	+0.22	+0.16
50	-0.18	-0.02	-0.16	+0.01	-0.16	-0.17	-0.01	-0.10
75	-0.32	-0.17	-0.30	-0.16		-0.31	-0.17	-0.24
90	-0.16		-0.15			-0.16		-0.16

is otherwise repeatable, the results of the tests can be used to develop a new, in-place reference for the instrument.

Deficiencies

There are several potential deficiencies in conducting acceptance tests on instruments under operating conditions.

- It may not be possible to adjust the process through complete range. Test five points around expected operating range, i.e., 60%, 70%, 60%, 50%, 60%.
- There may be no good reference test or test device. Testing may be performed using calibration equipment offered as an option by many manufacturers. A simulated primary element output is input to the transmitter. This type of testing assumes that the primary element is operating properly.
- There may be a lack of time, skilled test personnel, or money. Require that the manufacturer conduct the acceptance testing and witness testing. Schedule startup to allow time for testing.
- The manufacturer may not have included support services in the price. Ensure that specifications require manufacturers to provide testing services.
- Instrument placement may be such that testing is difficult. During the process design phase, verify layout of equipment and piping to allow in-place calibration and acceptance testing for all instrument installations.
- Testing errors may result in inconclusive or disputed test results (see below).

Effect of Testing Errors

Before accepting or rejecting an instrument, the accuracy of the test should be determined. The SAMA Standard requires the following:

> If accuracy of the reference measurement is one-tenth or less than that of the instrument under test, then reference test measurement errors may be ignored. If the reference accuracy is one-third or less, but greater than one-tenth, it must be considered in evaluating instrument accuracy. Reference measurements with an accuracy of greater than one-third the test instrument should be avoided.

Consider the following example for calibrating a doppler flow meter, the test instrument, upstream from a magmeter, the reference instrument.

Assume both meters have a range of 0–10 mgd and that the magmeter is accurate to ±0.5%. As indicated previously, the percent error is:

$$E\% = [(V_A - V_R)/V_R] \times 100$$

Where $E\%$ = percent error
            V_A = doppler reading
            V_R = magmeter reading (reference)

The accuracy of $E\%$ can be determined by applying differential calculus. The result is:

$$dE\% = [dV_A/V_R - (V_A/V_R^2)\, dV_R] \times 100$$

Where $dE\%$ = error of the calculated percent error, $E\%$
            dV_A = doppler reading accuracy
            dV_R = magmeter reading accuracy

Assume the following:

1. The magmeter flow transmitter reading is recorded on pen 1 of a strip chart and the doppler reading is recorded on pen 2 with some offset to allow readability.
2. The magmeter reads 5.0 mgd and the doppler reads 5.1.

The error, $E\%$, would be reported as +2%. $[(5.1 - 5.0)/5.0] \times 100$

Now assume the following:

1. The magmeter is accurate to $\pm 0.5\%$ or ± 0.025 mgd
2. We can read the strip chart to the nearest $\pm 0.25\%$ or ± 0.013 mgd

    $dV_A = \pm 0.013$
    $dV_R = \pm 0.013 \pm 0.025 = \pm 0.038$ under worst case conditions

The possible error in $dE\%$ is ± 0.26 to $\pm 0.78\%$ or about $\pm 1\%$ under worst case conditions. This implies that the actual error, $E\%$, could range from 1 to 3% at worst to between 1.5 to 2.5% at best.

When performing acceptance testing, you must consider the sources of error in both the reference and actual measurements. Sources of error include:

1. inaccurate reading strip charts or gauges
2. accuracy of reference meter or reference test
3. whether testing errors are cumulative or tend to cancel each other

DETERMINING INSTRUMENT REFERENCE VALUES

1. ANALYTICAL INSTRUMENTS

The following are general methods for determining analytical instrument reference values:

1. lab analysis
2. portable reference probe
3. test solution of known concentration

These methods vary in difficulty and accuracy. The method selected is determined by the type of meter being tested, the type of liquid being measured, and the resources available to conduct the test.

Important considerations for these methods are:

1. The lab sample must be representative of the process at the time test instrument readings are taken.
2. Reference probes must be accurately calibrated.
3. Test solutions must be representative of the process stream.

Chlorine Residual

Standard Methods lists seven methods for determination of chlorine residual. The amperometric method is recommended because it is not subject to interference from color, turbidity, iron, manganese, or nitrite nitrogen. The DPD methods are simpler for determining free chlorine than the amperometric titration method.

Commercially available chlorine residual analyzers use the amperometric, DPD, and iodometric methods. Chlorine in aqueous solution is not stable, and the chlorine content of samples will decrease rapidly. Chlorine determinations should be started immediately after sampling. Excessive light and sample agitation should be avoided.

Dissolved Oxygen

Dissolved oxygen reference values are best determined with a reference probe. A reference probe is calibrated in air and against laboratory prepared solutions of known dissolved oxygen concentration. Portable probes should be accurate to ± 0.1 mg/L. Laboratory iodometric tests to calibrate the probe are accurate to ± 0.05 mg/L.

pH

Two methods are available to determine pH reference values, titration and electrometric. For both methods, samples must be handled carefully to avoid loss of dissolved gases or to avoid entrainment of gases. The temperature of the sample should be the same as the process.

Titration

This method can only be used with samples that are not colored or turbid. Turbidity may obscure the color change at the end point. Standard Methods states that the accuracy is about ± 0.3 pH units.

Electrometric

Use a glass electrode laboratory probe with an accuracy of at least ± 0.02 pH units. Calibrate the probe against standard buffer solutions of known pH. Standard Methods states that under normal conditions, an accuracy of ± 0.1 pH unit can be achieved.

Suspended Solids and Turbidity

Standard Methods states the standard deviation of laboratory suspended solids tests, the reference, was 5.2 mg/L (33%) at 15 mg/L, 24 mg/L (10%) at 242 mg/L, and 13 mg/L (0.8%) at 1707 mg/L. It may be hard to reject a suspended solids analyzer which is not within ± 1% of the laboratory result, especially in the low solids ranges.

Suspended Solids (20,000 to 80,000 mg/L)

Total, fixed, and volatile solids in semisolid samples are suitable for determining a suspended solids reference value when the solids concentration is above 20,000 mg/L. Calcium carbonate may be used as a reference solution to test instrument accuracy.

Suspended Solids (500 to 20,000 mg/L)

Total suspended solids dried at 103–105°C test are suitable for determining a suspended solids reference value. Ferric hydroxide may be used as a reference solution to test instrument accuracy.

Suspended Solids (5 to 500 mg/L)

Total suspended solids dried at 103–105°C test are suitable for determining a suspended solids reference value.

Turbidity (0 to 40 NTU)

Turbidimeters are calibrated against standard reference suspensions of formazin polymer suspension. Analyze grab samples with a calibrated laboratory turbidimeter.

2. LIQUID FLOW INSTRUMENTS

The following are general methods for determining reference liquid flow values:

1. volumetric
2. comparison with a reference flow meter
3. dilution
4. salt velocity
5. velocity area

These methods vary in difficulty and in accuracy. The method selected will be determined by the type of meter being calibrated, the type of liquid being measured, and the resources available to conduct the test.

Volumetric Calibration

The feasibility of the volumetric calibration (fill/draw) method depends primarily on the availability of suitable tank space and connecting conduits. Important considerations for this method are:

1. Conduct the test using the process liquid to be measured under normal flow meter operation.
2. The potential accuracy of this method is high.
3. The tank should be regularly shaped so its volume can be calculated within acceptable limits of accuracy.
4. The tank volume should be large enough to provide a test run long enough to make start and finish timing errors negligible.
5. The change in liquid level in the tank should be enough so that starting and finishing depths can be measured without introducing significant error.
6. The flow rate should remain relatively constant during the test run.

Estimating the percentage error for this testing method should include an estimate of errors introduced due to physical measurements of tank volume, depth change, and the elapsed time of the test.

Comparison with a Reference Meter

A reference meter is a flow-measuring device whose performance characteristics can be referenced to published standards or to recommended practices acceptable to involved parties. Examples include:

1. standard venturi tubes and venturi nozzles (2,3,4)
2. orifice plates (2,3)
3. Parshall flumes (4,5)
4. thin plate weirs (5)

Numbers in parentheses are Bibliography numbers.

Important considerations regarding this method include:

1. The flow meter(s) used as reference devices must meet all requirements of accepted standard practices in fabrication, installation, and use. In most treatment plant applications, conformance to these requirements, especially for installation and use, is difficult.
2. Use of differential pressure type flow meters requires pressure differential measurement with a U-tube mercury manometer.
3. When standard weir or flume methods are used, a point gauge must be used for head measurement. Surface foam or scum, as well as difficulty in reading the point gauge, will affect the results.

Dilution Method

With the dilution method the flow rate is deduced from the dilution of measurable properties of tracer chemicals added to the flow in known amounts. The tracer can be injected either in a constant rate or in a one-shot plug. The constant-rate method is more suitable in treatment plant applications. Consult Bibliography numbers 6 and 7 for greater detail of this method.

In the constant-rate injection method, a tracer solution of accurately known concentration is injected upstream at a constant, accurately measurable rate. At a downstream distance sufficient to achieve complete mixing and a steady-state tracer concentration, the flow is sampled. The tracer concentration is determined and used to calculate the flow rate.

Important considerations regarding this method include:

1. The tracer property measured must be conservative. Rhodamine WT has been used successfully in raw sewage. Its behavior in sludges is not known.
2. Accurate measurement must be made of the rate at which the tracer is added, and the initial and final concentrations of the tracer.

3. A high sensitivity spectrophotometer is required for this method.
4. Accuracy is $\pm 2\%$.

Salt-Velocity Method

In the salt-velocity method, brine is injected suddenly at an upstream station in such a way that it rapidly becomes distributed across the pipe section. The time of passage of the salt pulse between two downstream stations is measured by conductivity-sensitive electrodes. The flow rate may then be determined if the volume of the conduit between the electrodes is accurately known. Consult Bibliography numbers 3 and 8 for details.

Important considerations regarding this method include:

1. The best attainable accuracy for this method is $\pm 1\%$.
2. The process liquid being measured must have a significantly smaller conductivity than the brine solution.
3. The method is not satisfactory for use with raw sewage or sewage sludges because of conductivity fluctuations. It is suitable for treated effluent and raw water.

Velocity-Area Method

This method is applied to a flow cross section by measuring a number of velocities over the section, each representative of the average velocity within an incremental area, and then summing the resulting velocity-area products. This method can be applied to both open and closed conduit flows, but it is more conveniently employed in accessible open channels. Consult Bibliography number 9 for further information.

Important considerations regarding this method include:

1. The individual velocity components can be measured by point velocity measuring instruments, e.g., current meters, pitot tubes, or by acoustic velocity meters that measure an average velocity component along a line path.
2. The point velocity meters are intrusive and may not work well with raw sewage and sewage sludges.
3. The velocity sampling requirements are lengthy and this method is suitable only where long periods of steady flow are available.
4. Accuracy can be as poor as $\pm 5\%$ of reading.

3. GAS FLOW INSTRUMENTS

The following are general methods for determining reference gas flow values:

1. volumetric
2. comparison with a reference flow meter
3. comparison with manufacturer's data

The methods vary in difficulty and in accuracy. The method selected is dependent on the process piping and type of gas.

Volumetric Calibration

The feasibility of the volumetric calibration method depends primarily on the availability of suitable tank space and connecting conduits.

Important considerations for this method are:

1. Conduct the test using the process gas to be measured under normal flow meter operation.
2. The potential accuracy of this method is high.
3. The tank should be regularly shaped so its volume can be calculated within acceptable limits of accuracy.
4. The tank volume should be large enough to provide a test run long enough to make start and finish timing errors negligible.
5. The change in pressure and temperature in the tank should be enough so that starting and finishing values can be measured without introducing significant error.
6. The flow rate should remain relatively constant during the test run.

Estimating the percentage error for this testing method should include an estimate of errors introduced due to measurements of tank volume, pressure and temperature change, and the elapsed time of the test.

Comparison with a Reference Meter

This test would be similar to that for liquid flow meters.

Comparison with Manufacturer's Data

If the gas feeds a burner, burner rating can be used to estimate the flow.

4. LEVEL INSTRUMENTS

The following are methods for determining the level reference value:

1. direct measurement
2. pressure

Direct Measurement

The feasibility of the direct measurement method depends on the accessibility to the process stream. Direct measurement is easiest for open tanks and channels.
Important considerations for this method are:

1. Use surveyor's instruments (level or metal tape). Accuracy can be ±0.15 cm (±0.005 ft).
2. Use a point gauge if the process stream is easily accessible. This type of instrument is very accurate, but is normally limited to laboratory use.

Pressure

Pressure can be used where direct measurement is not practical.
Important considerations for this method are:

1. The density of the liquid must be determined.
2. Use a manometer, either water or mercury, to obtain the most accurate pressure.

5. PRESSURE INSTRUMENTS

A water or mercury manometer is the most accurate method of obtaining a pressure reference value.
Important considerations for this method are:

1. Place taps so they are flush with the pipe inner surface and free of burrs.
2. Do not use piezometric rings in wastewater applications.
3. Route the manometer connecting tubing to avoid accumulation of gases or solids. Slope horizontal lines. Place bleed valves at high points.

PERFORMANCE MONITORING

In addition to carrying out acceptance and calibration testing, conduct performance monitoring on an ongoing basis. Performance monitoring can provide plant personnel with a quick indication of instrument performance by using a secondary instrument or analysis or calculation results which measures the process variable to an accuracy of 5–10% of actual.

Analytical Instruments

The easiest method for analytical instrument performance monitoring is to compare daily average readings with lab analysis of daily composite samples.

Flow Meters

Flow meter performance can be monitored by measuring pressure differences or by comparing flow against manufacturer-prepared rating curves and tables.

Measuring Pressure Difference

Measuring a pressure difference at a location where the flow rate has a unique and repeatable relationship to the primary flow meter provides a common and adequate means for monitoring meter performance. The pressure difference may be measured around any pressure differential-causing hydraulic element, with 90° pipe elbows being most commonly used.

This type of monitoring should be available prior to acceptance and calibration testing of the primary flow meter, so that a relationship between the pressure monitor and the flow rate indicated by the meter can be established.

Consider the following for the use, placement, and operation of pressure taps to monitor pressure difference.

1. Mount the taps so they are flush with the pipe inner surface and free of burrs.
2. Hole diameter is not extremely critical. Diameters from 0.5–1.0 cm (1/64–3/8 in.) for small to large pipes are usually adequate.
3. Install the taps along the axial center of pipe elbows. For other locations, install them both up- and downstream of the device producing the pressure difference.
4. Multiple taps with piezometric rings are not recommended for use with wastewater treatment process streams. Use single taps, placed upstream and downstream.
5. Locate single taps in horizontal or near-horizontal process lines in a

horizontal diametric plane to minimize gas and/or solids entry into the measurement lines.

Consider the following aspects of the differential pressure sensing device and connecting manometer tubing when setting up the monitoring system:

1. When U-tube manometers are used, the medium selected should be appropriate for the magnitude of the pressure difference expected, e.g., water-air for small differentials and water-mercury for large differentials so that a minimum deflection of 7.5 cm. (3 in.) should be maintained.
2. When commercial differential pressure cells are used, they should be calibrated frequently to a liquid column manometer.
3. Route manometer connecting tubing to avoid accumulation of gases and/or solids. Slope horizontal lines. Place bleed valves at high points.
4. The connecting tubing should be corrosion resistant with a nominal diameter of 1 cm (3/8 in.). Provide it with valving for periodic flushing.

Manufacturer-Prepared Rating Curves and Tables

Approximate flow measurements suitable for performance monitoring may also be obtained from manufacturer-prepared pump rating curves and flow vs angle of opening data for butterfly valves. These measurements would be subject to inaccuracies caused by installation and/or operational factors but nonetheless provide an easy means of ascertaining gross meter inaccuracies.

Level Instruments

In open tanks and channels, a measuring stick or rope can be used to periodically check the level instrument.

In closed tanks and wet wells, a site glass should be provided. Mount a tape measure next to the site glass for quick comparisons.

Pressure Instruments

Pressure gauges provide a quick means of monitoring pressure instrument performance. The pressure gauge should be lab-calibrated periodically.

BIBLIOGRAPHY

1. National Bureau of Standards, "Recommended Practice for the Use Of Electromagnetic Flow Meters in Wastewater Treatment Plants" (Cincinnati, OH: U.S. Environmental Protection Agency, Municipal Environmental Research Laboratory, August, 1980).
2. "Measurement of Fluid Flow by Means of Orifice Plates, Nozzles and Venturi Tubes Inserted in Circular Cross-Section Conduits Running Full," ISO/DIS 5167, draft revision of R781 (New York: International Standards Organization, 1976).
3. "Standard Methods of Flow Measurement of Water by the Venturi Meter Tube," ASTM D2458-69 (Philadelphia: American Society for Testing and Materials).
4. "Standard Method for Open Channel Flow Measurement of Industrial Water and Industrial Wastewater by the Parshall Flume," ASTM D1941-67 (Philadelphia: American Society for Testing and Materials).
5. "Methods of Measurement of Liquid Flow in Open Channels," Standard No. 2680-4A, Part 4A (London: British Standards Institution, 1965).
6. "Measurement of Water Flow in Closed Conduits – Tracer Methods, Part I: General," ISO No. 2975/1 (New York: International Standards Organization, 1974).
7. "Part II: Constant Rate Injection Method Using Non-radioactive Tracers," ISO No. 2975/2, (New York: International Standards Organization, 1974).
8. "Standards for Centrifugal, Rotary and Reciprocating Pumps," 12th ed. (New York: Hydraulic Institute).
9. "Liquid Flow Measurement in Open Channels – Velocity Area Methods," ISO 748 (New York: International Standards Organization, 1976).
10. "Standard Methods for the Examination of Water and Wastewater," 16th ed. (Washington, DC: American Public Health Association, 1985).
11. "Process Instrumentation and Control Systems Manual of Practice," OM-6 (Washington, DC: Water Pollution Control Federation, 1984).
12. "Test Protocol for Total Residual Chlorine Monitor," (Washington, DC: Instrument Testing Service, Inc.).
13. "Test Protocol for On-Line Dissolved Oxygen Analyzers for Wastewater Treatment Plant Applications" (Washington, DC: Instrument Testing Service, September 1986).
14. "Draft Test Protocol for On-Line Suspended Solids Analyzers" (Washington, DC: Instrument Testing Service, February 1987).
15. "Generic Test Methods for the Testing and Evaluation of Process Measurement and Control Instrumentation," SAMA Standard PMC 31.1-1980 (Washington, DC: Scientific Apparatus Makers Association).
16. "Statistical Process Control for Water Treatment Systems," *Chem. Eng.*, (May, 1987).

Chapter 8

Pumps for Variable Flow Service

This chapter explains some ways in which pumps are used to control flows. Where pertinent to control, principles of pump operation are included. Subjects such as sizing, mechanical details, and manufacturer's options are not included in this discussion. For further information on these subjects, see the Bibliography at the end of the chapter.

Three classes of pumps are discussed here: metering, positive displacement, and centrifugal. Although metering pumps use the positive displacement principles, they are considered separately because they form a major subgroup of positive displacement pumps.

METERING

Applications

In treatment plants, metering pumps are used to add polymer or other chemicals to clarifiers for aid in settling or for precipitation of pollutants, to add chemicals to boiler makeup water, to add chemicals for dewatering of sludges, to add acid or base for pH control, etc.

In an open-loop control system, the pump is usually either adjusted manually to the desired rate of addition or paced to the main process flow. Manual adjustments are recommended only for processes that operate at a constant or nearly constant flow. When addition is paced to the process flow, the addition rate is calculated as follows:

$$Q_A = (D \times Q)/(P \times 1000)$$

Where D = dosage, mg/L
                 Q = main process flow, L/min
                 Q_A = chemical addition rate, L/min
                 P = density of chemical g/L

For variable stroke pumps, the stroke length is:

$$L = Q_A/(S \times A)$$

Where S = pump speed, strokes/min
                 A = piston area, cm^2
                 L = stroke length, cm

And for variable speed pumps, the speed is:

$$S = Q_A/(L \times A)$$

Typically, the flow signal and the pump control signal are both 4–20 mAdc, and the dosage is set at an electronic ratio station. If both speed and stroke are variable, the practice is to pace pump speed and manually adjust stroke to set the dosage, thus eliminating the ratio station. In open-loop applications, errors in stroke and speed accuracy are not compensated by controller action.

Accuracy of stroke adjustment is typically stated to be ±1% of full capacity, with linearity and repeatability also ±1%. In terms of absolute error, stroke adjustment at constant discharge pressure would perform as follows:

% Full Stroke	Stroke Absolute Error	Pump Speed Absolute Error
100%	± 1%	0%
50%	± 2%	±1%
25%	± 4%	±1%
10%	±10%	±1%

Error is also introduced by a change in motor speed resulting from a change in load. Another ±1% absolute error is added to the stroke error for 50% stroke and less. During the life of a pump, error will increase due to wear on moving parts that allow hydraulic leaks and metered liquid leaks. Another source of error is entrainment or buildup of gas that causes some piston displacement to be negated by compressibility of the gas. Mass flow rates will be affected by changes in metered liquid density (e.g., due to temperature changes).

Closed-loop systems involve a process measurement downstream of the chemical addition point. A controller uses this measurement to vary the metering rate to match the process setpoint. Errors in speed or stroke adjustment are corrected by the controller. In closed-loop applications the process requirements for turndown of chemical addition can influence the choice of variable drive. Variable stroke pumps can turndown 100 to 0% of rated pump output, although stroke accuracy would normally limit turndown to 10:1.

Variable speed pumps, on the other hand, are typically limited by the type of drive from 3:1 turndowns to about 10:1 turndowns. Examples of open-loop and closed-loop applications are shown in Figure 8.1.

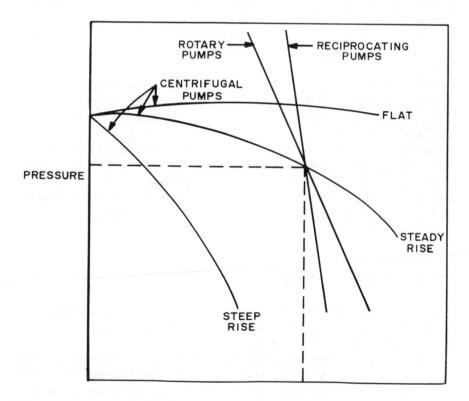

Figure 8.1. Typical pump curves, at constant speed.

Principle of Operation

Metering pumps are positive displacement-type pumps of either the reciprocating or rotary type. Reciprocating metering pumps have two general methods to vary pump output: variable speed or variable stroke. Stroke on a variable speed pump is adjusted by either varying the crank travel directly (amplitude modulation) or by varying the amount of fixed-crank travel transmitted to the piston.

Amplitude modulation can also be achieved by using a slider-crank in which the length of a pivot arm (or eccentric) is adjusted. Adjusting the pivot arm to zero length results in zero piston travel. Design of slider-crank mechanisms vary by manufacturer. An example is shown in Figure 8.2.

Another method to modulate amplitude is by the shift-ring drive in which the piston rotates in a ring that can be positioned. Fixed-crank stroke adjustments use a lost-motion drive that limits the piston or diaphragm travel. Motion can be lost in the transfer from crank to piston as shown in Figure 8.3.

The piston reciprocates as it follows the eccentric cam until the piston is stopped by the stroke adjustment pin. Piston travel is resumed when the eccentric rotates enough to pass the position of the piston stop. Another lost-motion drive is shown in Figure 8.4. Here motion is lost between the piston and the diaphragm by allowing some of the hydraulic fluid to escape to a reservoir.

Variable speed drives on a metering pump can be eddy-current, Silicon Controlled Rectifier (SCR), variable frequency, or belt drives using conical pulleys to vary the ratio of drive to pump speed. Variable speed drives are discussed in Chapter 9.

Metering pumps can have variable speed, variable stroke, or both variable speed and variable stroke capabilities. The choice of variable drive depends on the pump's application. Another method of metering uses a recycle valve to vary pump output. Variable rate pumps are used in automatic control to eliminate the cost and maintenance of a control valve. The variable rate pumps also save energy.

POSITIVE DISPLACEMENT

Positive displacement (p.d.) pumps are usually diaphragm, piston, or progressive cavity types. These pumps are used to pump liquids with high solids content such as thickened sludges.

Diaphragm and piston pumps are large-scale versions of the metering pumps discussed previously. Variable speed and variable stroke control capabilities also apply here. Progressive cavity pumps do not have variable stroke capability because the rotor and stator are not adjustable.

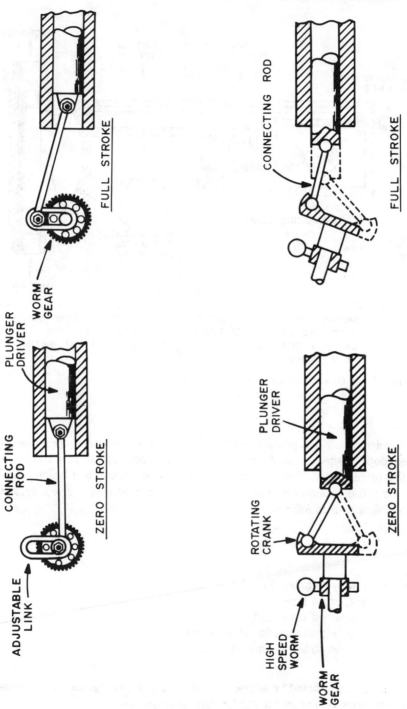

Figure 8.2.   Slider-crank stroke adjustment.

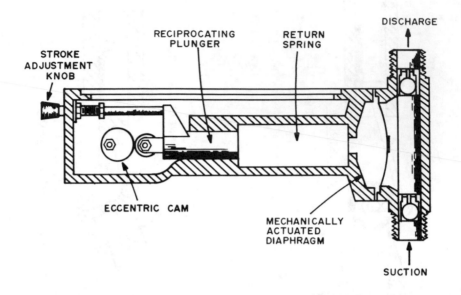

MECHANICAL LOST-MOTION DRIVE CHANGES THE DISCHARGE
FLOWRATE BY VARYING THE PLUNGER RETURN POSITION.
CRANK ECCENTRICITY REMAINS CONSTANT OVER THE ENTIRE
FLOW RANGE.

Figure 8.3.   Eccentric cam lost-motion stroke adjustment.

Control methods for positive displacement pumps are either open-loop or closed-loop. A typical open-loop system is sludge withdrawal from a primary clarifier or thickener. Constant rate p.d. pumps are controlled by an on/off timer basis (i.e., open-loop). On time and off time are adjusted to give a desired average flow rate as follows:

$$Q = [t_r / (t_r + t_o)]SV$$

Where Q = pumped flow, L/min
         t_r = on time, min
         t_o = off time, min
         S = pump speed, strokes/min
         V = stroke volume, L/stroke

On and off time should be adjusted if necessary to prevent excessive wear and overheating on the motor starter and windings.

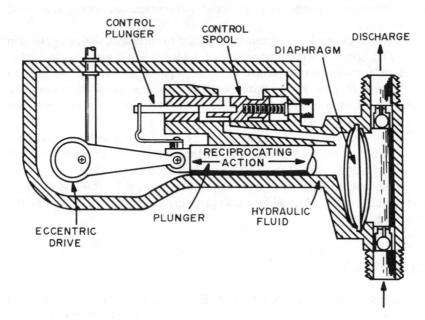

A PORTION OF HYDRAULIC FLUID IS PERMITTED TO ESCAPE
THROUGH A BYPASS VALVE WITH EACH STROKE, THEREBY
CHANGING THE EFFECTIVE STROKE—LENGTH OF THE PLUNGER.
NOTE THE BALANCED-DIAPHRAGM LIQUID END ASSOCIATED
WITH THIS DESIGN

Figure 8.4.   Hydraulic lost-motion stroke adjustment.

Other control strategies include:

1. Set on time in terms of volume of clarifier influent flow. This needs a clarifier influent flow meter.
2. Stop pump based on time or low density cutoff, whichever occurs first. This needs a suspended solids analyzer.
3. Start pump at high blanket level and stop at low blanket level. This requires two blanket level detectors.
4. A combination of any of the above options.

A variable rate pump can be operated with the same control method, but the control can be augmented to take advantage of variable speed properties. For example, pump rate may be regulated in closed-loop control to maintain measured suspended solids concentration at a setpoint.

In some cases, the receiving process may be upset by large variations in flow. A variable rate pump can be set for steady, continuous operation to avoid this condition.

Variable rate pumps have limitations in regard to settling of sludge or solids in piping. This occurs at low liquid velocities, under 0.8 to 1 m/s (2.5 to 3 ft/s). Therefore, for variable rate pumps, the system piping needs to be designed to have this velocity at maximum turndown. The pipe size calculated to meet this criterion may be too small based on other system requirements, in which case variable rate pumps may be unsuitable.

Output of p.d. pumps can be modulated by using a recycle line and control valve. Throttling valves are not used on the suction side because they can cause cavitation problems. On the discharge side they can cause overpressurization. Protection from overpressurization is essential for p.d. pumps and is usually achieved by a high pressure cutout switch located on the pump discharge.

CENTRIFUGAL

Centrifugal pumps are characterized by the output pressure (head) produced at various flow rates. Typical centrifugal pump curves are shown in Figure 8.5. Maximum head occurs either when the pump is deadheaded, or when drooping at a low flow occurs. As the flow increases, the head falls off. Compared to positive displacement pumps, a centrifugal pump's capacity is very sensitive to changes in discharge pressure. Degree of sensitivity varies from flat response to steep response, with most pump designs falling in between with steady rise characteristics as shown in Figure 8.1.

Size a centrifugal pump by matching the pump characteristics to the process. An example is shown in Figure 8.5. The pump will operate at the pressure and flow intersection with the system head loss curve. In the figure, system curve 1 represents maximum flow required by the process plus a safety factor; curve 2 represents a throttled state at minimum flow. A good match occurs when the pump efficiency is within the operating flow range, near the average flow. System curves or operating design data are calculated from fluid dynamic principles, while pump curves are available from manufacturer's literature.

Even though a pump has a curve that meets the processes requirements at high efficiency, it may not be the best pump for the application. The designer must also consider average and maximum horsepower (brake-horsepower) requirements, net positive suction head, impeller design, number of stages, volute design, diffuser design, and mountings. These topics are beyond the scope of this chapter, but are presented in the Bibliography.

The most common flow control method for centrifugal pumps is with a throttling valve. In design selection, a portion of the pump head at maximum flow is used for loss at the throttling valve. This portion may range from as low

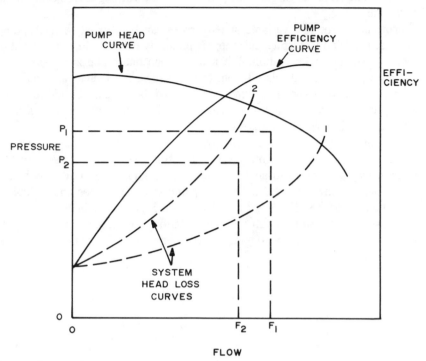

Figure 8.5. Sizing curves for centrifugal pumps.

as 20% up to 33%. It is set by desired throttling valve gain characteristics, desired safety factors, and energy conservation.

For small process flows, the portion is often toward the higher percentage because energy conservation is not as important as gain and safety factors. The opposite is true for large process flows. Here, lower percentages are favored because of lower energy costs. Thus, pump sizing is strongly affected by the valve size. This is discussed in more detail in the chapter on control valves.

The output of a centrifugal pump can be controlled with a recycle valve. In this case, valve sizing and valve head loss do not affect pump sizing. To allow total recycle, all the pump head is lost or dissipated at the valve.

Pump output can also be changed by varying pump speed. A variable speed drive, such as variable frequency drive, eddy current coupling, or SCR drive, is used to control pump speed. These and other variable speed drives are discussed in Chapter 9. These drives have lower energy requirements than control valves. Due to inefficiencies of a drive, energy savings do not occur until average pump operation is under a certain percentage of pump speed. Several sources estimate this break-even point to be about 80 to 90% of full speed. A comparison of variable speed drives with throttling valves is shown in Figure 8.6.

Throttling valves and variable speed drives can be applied to multi-pump systems. In addition, starting and stopping pumps is a way to alter the output of a booster station or lift station. When more than one pump is in operation, station output can be calculated from the curves of the individual pumps. In general, pumps in parallel need similar curves and their capacities add together which produces a longer, flatter curve. Pumps in series heads add together, which results in a higher, steeper curve.

For example, consider a two-pump system, with identical pumps in parallel. The pump station output is controlled by a throttling valve. A flow meter on the station discharge is used to shut off one of the pumps when flow falls below 40% of the station's capacity, and start a pump when flow rises above 50% of the station's capacity. The system is shown on Figure 8.7. In this example, head losses at the throttling valve are a substantial portion of the system head. Could the use of variable speed pumps save energy somewhere below 100% of station output?

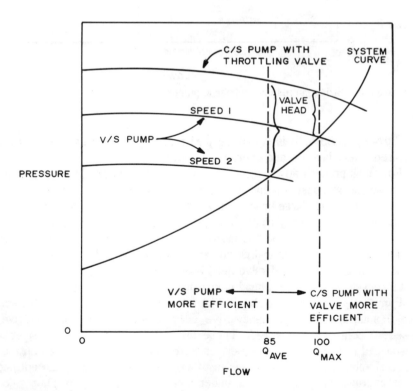

Figure 8.6. Variable speed versus throttling of centrifugal pumps.

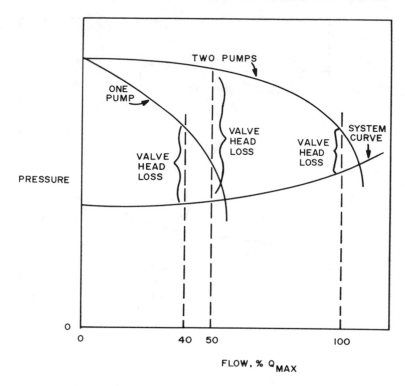

Figure 8.7. Two identical pumps in parallel with throttling valve.

Now look at the same pump curves where the pumps are variable speed. Assume that pump turndown is limited to 70% of maximum speed by the construction of the variable speed drive. Speed is related to pump performance approximately as follows:

Flow capacity = (speed/max speed) × maximum capacity
Head = (speed/max speed)^2 × head at maximum speed
Horsepower = (speed/max speed)^3 × horsepower at maximum speed

Therefore, the turndown limit on capacity is from 100 to 70% of maximum capacity. This system is shown in Figure 8.8. Using just variable speed and start/stop control is not enough to provide a continuous range of control from one pump at minimum speed to two pumps at maximum speed. A control gap occurs from 50 to 70% of maximum station output. Either a recycle valve or throttling valve is needed to eliminate this gap. Energy savings should be expected if the average flow is less than about 80 to 90% of maximum.

When used for controlling wet well level in a lift station, centrifugal pumps

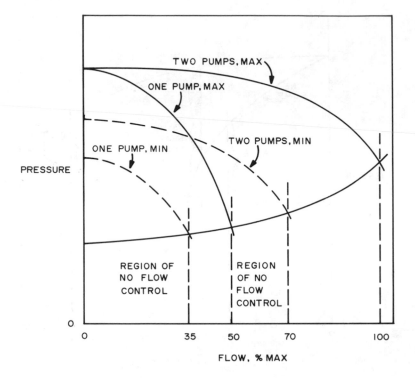

Figure 8.8. Two identical variable speed pumps in parallel.

are normally constant speed. The pumps are started and stopped at selected level trip points. Some booster stations use this control method with pressure used to provide trip points. Usually tight pressure control is not needed, and pressure is allowed to swing between limits.

Variable rate pumping systems are more common in treatment processes. In this case, level, flow, or pressure must be maintained within narrow limits.

BIBLIOGRAPHY

1. Poynton, J.P., "Basics of Reciprocating Metering Pumps," *Chem. Eng.* (May 21, 1979), p. 156.
2. Bristol, J. M., "Diaphragm Metering Pumps," *Chem. Eng.* (September 21, 1981), p. 124.
3. Coughlin, J.L., "Control Valves and Pumps: Partners in Control," *Instruments and Control Systems* (January, 1983), p. 41.
4. Hilsdon, C. W., "Using Pump Curves and Valving Techniques for Efficient Pumping," *Research and Technology* (March, 1982), p. 157.

5. "Energy Evaluation in Centrifugal Pump Selection," *Plant Eng.*(April 5, 1979), p. 71.
6. DeSantis, G. J., "How to Select a Centrifugal Pump," *Chem. Eng.* (November 22, 1976), p. 163.
7. Burris, B.E., "Energy Conservation for Existing Wastewater Treatment Plants," *Journal WPCF* 53(5) (May, 1981), p. 537.
8. Yedidiah, S., "Make Pumps with Drooping Curves Your Servants, Not Your Enemies," *Power* (March, 1982), p. 51.
9. Zell, B., "Pumps and Pumping Systems: Specifying to Save Energy," *Specifying Engineer* (October, 1981), p. 117.

Chapter 9

Variable Speed Drives

Variable speed drives are used for pumping or other mechanical functions that cannot be properly accomplished with a reasonable number of constant speed units. Variable speed drives also offer increased flexibility in control. Improved system efficiency can also be obtained where periodic changes in the demand allow operating at reduced horsepower to save energy.

MAGNETIC COUPLING

Application

A magnetic coupling variable-speed drive system uses a standard constant-speed induction motor to drive a coupling (clutch) that has an adjustable output speed. This drive offers good flexibility, with such options as braking and accurate speed control by a feedback signal produced by a tachometer generator on the output shaft.

Characteristics of a magnetic coupling are as follows:

1. It uses standard constant speed ac squirrel-cage induction motor.
2. A wide range of 10 to 750 kW (15 to 1000 hp) is available.

3. Cost is minimal, using simple variable speed drive.
4. It is suitable for variable torque pumping or fan loads, but not suitable for conveyors and piston pumps that are subject to heavy vibrations.
5. Low efficiency is achieved at reduced speeds. At low speed a large amount of power is dissipated as heat in the magnetic coupling. This requires cooling the magnetic coupling by air or liquid. Don't use magnetic couplings at speeds below 75% of nominal speed for an extended time because of lowered efficiency.
6. Water cooling to improve heat dissipation is generally required above 300 kW (400 hp).
7. Magnetic coupling reliability is good, and minimal maintenance is normally required. It has brushes, but unlike dc motors and wound rotor motors, they are on the magnetic coupling and carry only small currents.
8. Potential problems exist in maintaining proper alignment between the motor, magnetic coupling, and pump.
9. Additional floor space (for horizontal drives) or additional ceiling height (for vertical drives) is required, as compared to most other variable speed drives.

Principle of Operation

Magnetic coupling variable-speed drives use a constant-speed induction motor to drive a ferrous metal ring that rotates around a dc-excited magnetic rotor connected to the load. The magnetic rotor's dc current is increased or decreased to vary the degree of coupling force generated. Output speed of the drive is a function of the strength of the rotor magnetic field and the load on the output shaft. Output speed is automatically controlled by a speed controller that compares the output speed signal from the speed transmitter to a setpoint and adjusts the amount of dc current to the magnetic rotor to maintain the desired speed (see Figure 9.1).

Speed Control

Control for remote speed can be either by increase/decrease pushbuttons or by an analog signal (typically 4–20 mAdc). When increase/decrease control is desired, a motorized potentiometer or solid state ramp generator is required to produce an analog signal. Controllability is good, and accurate speed control can be obtained.

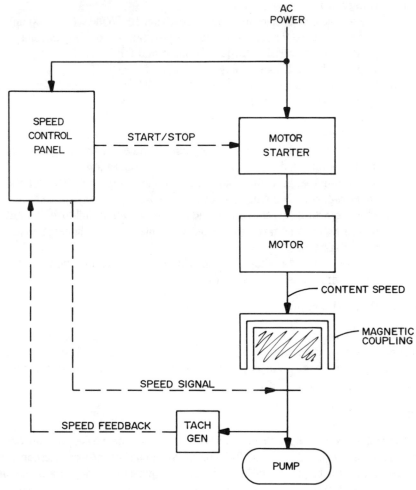

Figure 9.1. Magnetic coupling functional block diagram.

LIQUID RHEOSTAT

Application

A liquid rheostat variable-speed drive system uses a wound rotor motor. The motor speed is adjusted by changing the rotor current as determined by the depth in which the rheostat plates are submerged in liquid.

Characteristics of a liquid rheostat are as follows:

1. It uses wound rotor motor.
2. It is available in all standard motor sizes from 10–750 kW (15–1000 hp).
3. It is not suitable for constant torque loads, but it provides reasonably good drive for variable torque pumping and fan loads.
4. In general, a 2:1 speed range is available.
5. Overall efficiencies range from about 85% at full speed down to approximately 45% at half speed.
6. The power factor is poor at low speed.
7. A heat exchanger, with circulating pump, is required to dissipate from the liquid in the rheostat the heat produced by the electrical slip in the motor used to obtain variable speed operation.
8. Maximum speed, with minimum resistance between plates of the liquid rheostat, is about 95% of full speed.
9. Motor starting current can be reduced to less than full-load current by having maximum resistance between plates of the liquid rheostat during startup.
10. Problems with brushes and slip rings on the wound rotor motor occur with moderate frequency.

Principle of Operation

Electrical resistance between the liquid rheostat plates determines the wound rotor motor current which regulates the motor speed. The motor speed is adjusted by a change in the depth of the liquid in which the rheostat plates are submerged (see Figure 9.2).

Speed Control

Control for remote speed can be either by increase/decrease pushbuttons or by an pneumatic signal (typically 3–15 psig). When increase/decrease control is desired, a motor to pneumatic transducer is required to produce a pneumatic signal. If 4–20 mAdc control circuits are being used, a current to pneumatic transducer is required.

Controllability is generally good, but accurate speed control is difficult to obtain due to the slow response inherent with the pneumatic interaction required to change the submergence of the rheostat plates in the liquid.

VARIABLE FREQUENCY

Application

A variable frequency drive (VFD) system consists of an induction motor where both the voltage and frequency supply are controlled by an electrical inverter to adjust the motor's speed.

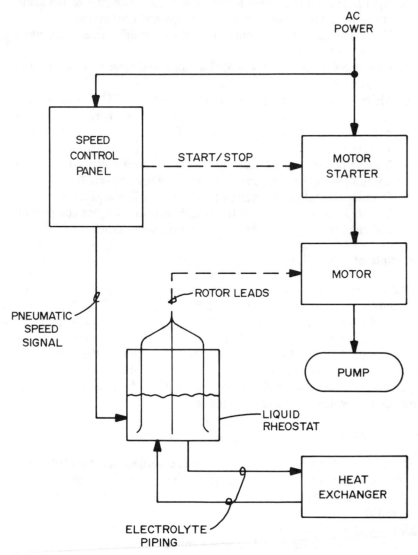

Figure 9.2.   Liquid rheostat functional block diagram.

Characteristics of a VFD system are as follows:

1. Other than the addition of thermostats in two phases of the stator winding, a standard squirrel-cage ac induction motor can be used.
2. All standard motor sizes are available from 4–400 kW (5–500 hp).

3. A VFD system is suitable for variable torque pumping or fan loads, and is normally suitable for piston pumps and conveyors.
4. Continuous operation at constant torque is available over a 3:1 speed range.
5. Overall efficiencies are about 83% at full speed down to approximately 75% at half speed.
6. Multiple motors can operate off one common VFD simultaneously. Also, one VFD can control more than one motor where it is switched between motors to give a combination fixed-speed and variable-speed system.
7. The VFD can convert existing constant speed motors to variable speed operation when retrofitting existing installations.
8. Starting current can be limited to less than full-load current.
9. The VFD requires little maintenance. However, complex components and circuitry require an expert technician when problems do occur.

Principle of Operation

The variable speed controller/power converter changes constant frequency, constant voltage line power to variable frequency, variable voltage power to vary the drive motor's speed. Raising the frequency of the power applied to the drive motor increases its speed. Lowering the frequency decreases its speed. Voltage applied to the drive motor is adjusted to control the motor's output power. The variable speed controller compares the drive motor's speed to an adjustable setpoint value and outputs the required frequency and voltage to maintain the desired speed (see Figure 9.3).

Speed Control

Remote speed control can be either by increase/decrease pushbuttons or by an analog signal (typically 4–20 mAdc). When increase/decrease control is desired, either a motorized potentiometer or solid-state ramp generator is required to produce an analog signal, or the specifications should require the ability of the VFD to handle momentary increase/decrease inputs. Controllability is excellent, and accurate speed control can be maintained.

VARIABLE PULLEY

Application

Electromechanical variable speed pulley belt drives use a variable sheave ratio principle to change the speed of the final element.
Characteristics of a variable pulley are as follows:

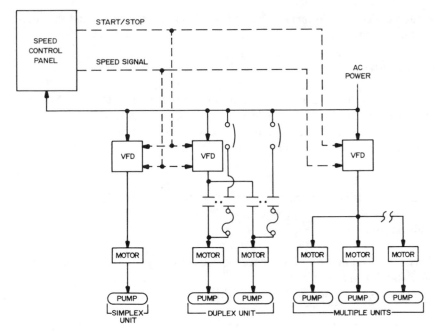

Figure 9.3. Variable frequency drive functional diagram.

1. It is available in all standard motor sizes from 0.2–75 kW (1/4–100 hp), but is normally used only in the lower ratings of 0.4–20 kW (1/2–25 hp).
2. This is a constant torque drive.
3. In general, a 10:1 speed range is available.
4. Overall efficiencies are about 70% at maximum speed down to approximately 45% at half speed.
5. Additional floor space or ceiling height is required as compared to most other drives.
6. Belt life is typically 18 months.

Principle of Operation

Variable pulley or electromechanical drives use a constant speed motor coupled to a variable belt drive system for speed control. Output speed is varied by changing the diameter of the drive sheave or pulley. Drive sheave diameter is adjusted using a reversing gear motor connected mechanically to the drive system. When the gear motor operates in one direction, the sheave diameter is increased and the drive speed increases. When the gear motor operates in the other direction, the sheave diameter decreases and the drive speed decreases.

Speed Control

Remote speed control is normally by increase/decrease pushbuttons. Potential operational problems holding a speed setting can be numerous due to belt wear, belt stretching, and groove-worn pulleys. Small stepless speed adjustments are difficult to obtain (and maintain).

DIRECT CURRENT (SCR)

Application

Variable speed control of dc motors is achieved by varying the armature or field voltage to the motor, or both.

Characteristics of dc variable speed drives are:

1. Direct current motors are considerably more expensive than comparable ac motors.
2. Drives are available from 4–100 kW (5–150 hp).
3. DC variable speed drives can be used for variable torque operation and are usually suitable for piston pumps and conveyors.
4. In general, a speed range of 60–100% is available.
5. Reduced speed efficiency is very good.
6. The dc motor has a commutator and brushes, both potential maintenance problems.
7. Commutators on dc motors can cause problems if they are not well ventilated and properly maintained, or if they are subject to a corrosive environment.
8. Drive requires little maintenance, but complex components and circuitry require an expert technician when problems do occur.
9. Starting current can be limited to less than full-load current.

Principle of Operation

A silicon-controlled rectifier (SCR) drive is a variable speed drive that is based on the use of a dc motor. Such SCR drives include a speed controller that uses an SCR-controlled bridge circuit to rectify constant voltage ac line power to variable dc voltage. Variable dc voltage is applied to the dc motor to regulate its speed. As the applied voltage increases, the motor speed increases. As the voltage decreases, the speed decreases.

SCR speed controllers accept an input signal representative of the desired motor speed and provide a dc output sufficient to operate the motor at the required speed.

Speed Control

Remote speed control can be either by increase/decrease pushbuttons or by an analog signal (typically 4–20 mAdc). When increase/decrease control is desired, either a motorized potentiometer or solid state ramp generator is required to produce an analog signal or the specifications should require the ability of the VFD to handle momentary increase/decrease inputs. Controllability is good, and accurate speed control can be maintained.

Chapter 10

Control Valves for Modulating Service

Control valves for modulating service are used in all water and wastewater treatment processes. Excluded from this discussion are valves used in isolation or routing service in two states: open or closed. Commonly used control valves are compared in Table 10.1.

VALVE TYPES

Control valves (modulating service is implied in this term) are made in a wide variety of designs. Five designs are discussed here to make you aware of some of the features and problems encountered in using control valves.

Ball Valves

Valves are classified as either "rotary" or "linear." In a rotary valve, the ball, disk, or plug is rotated to open or close the flow stream. A linear valve lifts the gate, disk, or plug up or down to open or close the flow stream. Ball valves are rotary. Flow goes through a port in the ball. To shut off flow, the ball is rotated until the port is closed. The ball may be a complete sphere, full ball, as shown in

Table 10.1. Comparison of Control Valves

Valve Type		Cost
Ball (rotary)	- High pressure recovery - Tight shutoff - Equal percentage flow characteristic - Available with flangeless connection - 1–24 in. sizes	Low
Butterfly (rotary)	- High pressure recovery - Equal percentage flow characteristic - Good shutoff - Usually a flangeless connection - 2–144 in. sizes	Lowest
Gate (linear)	- Not usually used for modulating service	
Globe (linear)	- Low pressure recovery - Choice of flow characteristics - Good shutoff - Usually small size applications - 1/2–16 in. sizes	High
Eccentric plug (rotary)	- Equal percentage and linear flow characteristics available - Good shutoff - Moderate pressure recovery - 2–24 in. sizes	Moderate

Figure 10.1, or a partial sphere. Standard port diameters range from 80% of inside pipe diameter to 100% or full port.

Ball valves have high friction resistance to rotation due to valve body and trim contact with the ball surface. This gives a tight shutoff at the expense of needing larger actuators. In some designs the ball is allowed to "float" in its seat so that line pressure will assist keeping a tight shutoff by pushing the ball against the downstream seal ring. This freedom of movement introduces a deadband between the actuator and the ball. The deadband will make this design unusable for many control applications. For control valves, a rigid ball to actuator connection is preferred.

Butterfly Valves

Butterfly valves are rotary type valves that use an axially pivoted disk to restrict or open the flow stream. The disk may be flat or contoured in shape and mounted to the stem (pivot) in several ways. An example is profiled in Figure 10.2.

One problem with butterfly valves is obtaining a tight shutoff. The seal must be made along the entire inner circumference of the valve body. Upper and

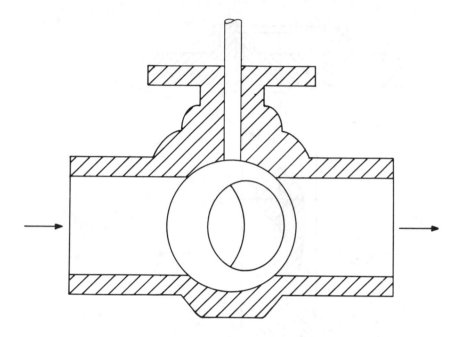

Figure 10.1.   Ball valve.

lower half seats must be matched and the pivot point must be sealed. Liners and seals used for good shutoff increase the opening and closing torque requirements of the actuator.

Another problem encountered in butterfly operation is torque applied by the fluid. During rotation of the disk, torque reaches a maximum at about 70 degrees (from full open). This is shown in Figure 10.3. In some applications this peak may exceed the opening or closing torques. To overcome this problem, disks are contoured for low torque in the near closed positions.

Gate Valves

Gate valves are linear valves in which the gate's disk is raised or lowered past a port by the actuator. The disk is a flat or wedged-shaped plate. The valve may have one or several ports for flow which can be sealed by the plate. A gate valve is shown in Figure 10.4. In knife-gate valves, plates are made with a sharp edge for service with solids-bearing streams including dry solids. Gate valves are not usually used for modulating service.

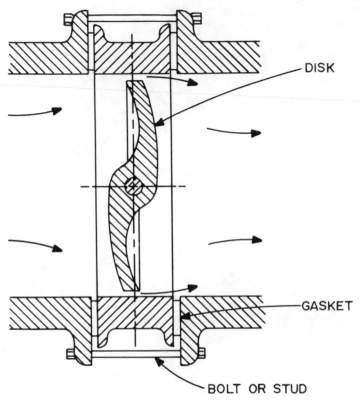

Figure 10.2.   Butterfly valve, swing-through type with flangeless pipe connection.

Globe Valves

Globe valves are linear valves in which the plug moves up or down into a port. Some valves have two sets of plugs and ports; these are called double-ported. In order to reduce stem size and to obtain better seating of the plug on the port, the stem or plug is mechanically guided. An example of a globe valve is shown in Figure 10.5.

Actuator size can also be reduced for double-ported valves with balanced flow. The ports and plugs are aligned such that the flowing stream tends to close one port and open the other, thus balancing the fluid forces.

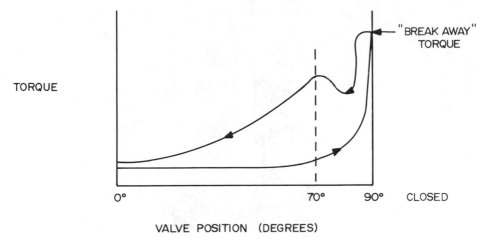

TORQUE

"BREAK AWAY"
TORQUE

0°   70°   90°   CLOSED

VALVE POSITION (DEGREES)

Figure 10.3. Butterfly torque.

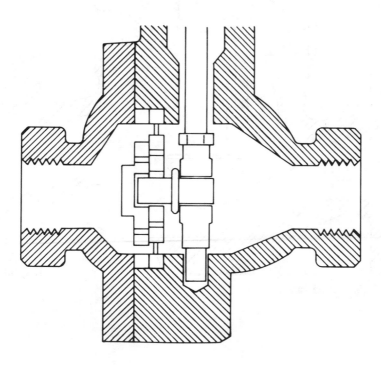

Figure 10.4. Gate valve, multi-orifice.

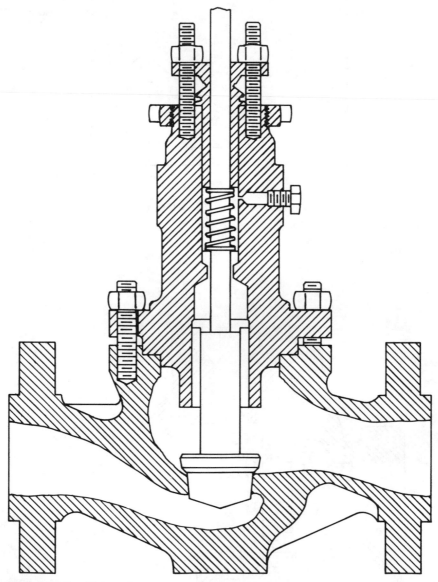

Figure 10.5. Globe valve.

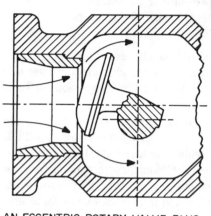

AN ECCENTRIC ROTARY VALVE PLUG

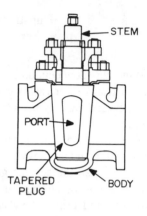

TRADITIONAL PLUG VALVE

Figure 10.6. Plug valves.

Plug Valves

Plug valves are rotary-type valves with a conical or cylindrical shaped plug which has an orifice. Rotating the plug at a right angle to flow causes tight shutoff while full flow occurs when the orifice of the plug is parallel to the flow axis. A simple plug valve is shown in Figure 10.6. In the same figure is a popular version of the plug valve, the eccentric spherical plug or "camflex" valve. Just as a traditional plug valve is similar to a traditional ball valve, the eccentric spherical plug is similar to the segmented ball valve. Both newer designs provide large flow for the valve size and high pressure recovery, yet have reduced friction through most of the valve travel.

VALVE SIZING

Control valves are rated by a "C_v" factor which is the amount of water at 15°C (60°F) or any liquid with a specific gravity of 1.0, that goes through a full open valve at a pressure differential of 6.89 kPa (1 psi). The C_v factor is found in the general flow equation below.

$$Q = C_v(DP/S.G.)^{1/2}$$

Where: Q = flow, gpm
C_v = valve flow coefficient gpm/(psi)$^{1/2}$
DP = pressure differential, psia = p_1-p_2
S.G. = specific gravity

p_1 = upstream absolute pressure (2 diameters)

p_2 = downstream absolute pressure (6 diameters)

While this flow equation is not restricted to English units of measure, the C_v values available for valves in this country are only for English units. To use kPa for pressure multiply manufacturer's English C_v values by conversions in Table 10.2.

The metric conversion factors also convert to a specific gravity at 4°C. For most water and wastewater treatment processes the liquid is water at 5–25°C (41–77°F), so the specific gravity factor will cause less than 0.5% correction and will not be discussed further here. Additional correction factors for viscosity and critical flow and compressibility for gas flow may be needed. These factors and how to apply them are found in manufacturer's literature. For a complete explanation of valve sizing see the ISA textbooks referenced at the end of this chapter.

In sizing a valve, usually the range of flows is known. A safety factor of 110% of maximum flow, or 130% of average flow is commonly used in the sizing calculations. Another way to add a safety factor is to make calculations using 100% of maximum flow, but selecting a valve that will do the job at 90% of its rated capacity (i.e., 0.9 C_v). Once the pressure drop over the valve is determined, the valve size can be calculated by rearranging the above equation. Traditional sizing rules for control valves in pump systems call for 33% of system pressure loss to be at the valve. This and other rules of thumb are shown in Figure 10.7. The lost energy is not wasted, but is used to reduce flows to the control setpoint.

This rule of thumb is not particularly energy efficient and should be replaced for large-sized flow streams where the energy cost is significant. Allotting less head loss to the control valve will tend to increase the valve size up to full line size, and will restrict the choice of valves to those with high capacity (e.g., butterfly or ball). Reducing head loss at the valve calls for more careful design than otherwise, because there is less safety factor. The risk is that the valve will be oversized for the application. In a severe case, the valve may have to operate nearly closed to achieve the setpoint flow, which may make satisfactory flow control impossible.

Table 10.2. C_v Conversion Factors

C_v × Conversion Factor		Flow in Units
1.44	=	Lpm
0.0862	=	m^3/h
0.00144	=	m^3/m
0.0000239	=	m^3/s

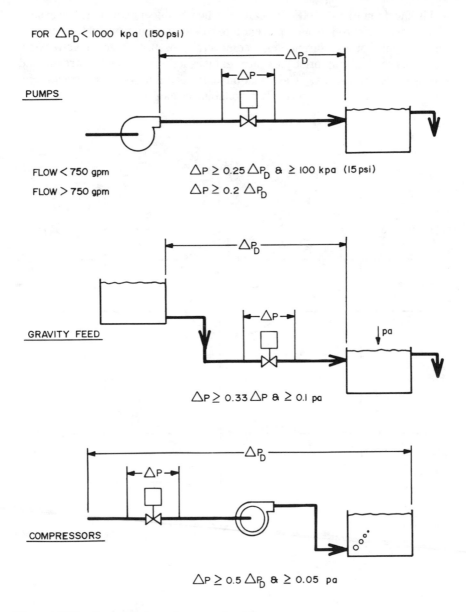

FOR $\triangle P_D < 1000$ kpa (150 psi)

PUMPS

FLOW $<$ 750 gpm $\triangle P \geq 0.25 \triangle P_D$ & ≥ 100 kpa (15 psi)
FLOW $>$ 750 gpm $\triangle P \geq 0.2 \triangle P_D$

GRAVITY FEED

$\triangle P \geq 0.33 \triangle P$ & ≥ 0.1 pa

COMPRESSORS

$\triangle P \geq 0.5 \triangle P_D$ & ≥ 0.05 pa

Figure 10.7. Pressure drops for control valves.

Pressure differential or drop across a valve is proportional to the square of flow. The pressure profile across a valve is similar to an orifice or venturi, and is shown in Figure 10.8. Similarly the flow is dependent on the pressure difference measured from upstream to the vena contracta. Due to the impracticality of measuring vena contracta pressure, control valve flow equations and sizing principles are established based on upstream and downstream pressures. The relationship between vena contracta and downstream pressure is normally a constant proportion expressed as:

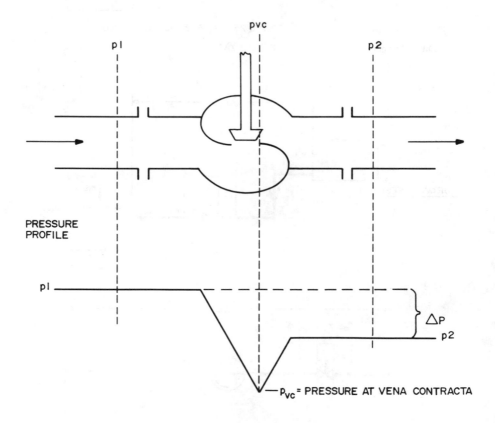

$$\Delta P = F_L^2 \, (P_I - P_{vc})$$

F_L IS THE PRESSURE RECOVERY FACTOR

Figure 10.8. Pressure drop across a valve.

$$DP = F_L^2(P_1 - P_{vc})$$

Where: DP = Difference between upstream and downstream pressures, kPa (psi)

P_1 = Upstream pressure, kPa (psi)

P_{vc} = Pressure at vena contracta, kPa (psi)

F_L = Pressure recovery factor, unitless

This relationship does not hold during conditions of cavitation or flashing in liquids or at sonic velocities in the vena contracta in gases. In these cases choked or partially choked flow occurs. When sizing valves for low head loss, choked flow and related phenomena are not likely problems except where the vapor pressure of a liquid approaches the vena contracta pressure as in hot water, for example. See the Bibliography for further information.

The recovery factor, F_L, is combined with other factors to form the C_v numbers reported in the manufacturer's literature. Therefore, the pressure difference of most use in sizing valves is the permanent pressure loss, DP.

The second consideration is valve gain and characteristic. Valve gain is defined as the change in flow caused by a change in valve position and may be dependent on valve position. The way in which valve gain changes with valve position falls into three basic characteristics:

1. Equal percentage—equal changes of valve position cause equal percentage changes in flow
2. Linear—flow changes linearly with valve position
3. Quick opening—flow changes rapidly with valve position at low valve position, but only slightly at higher valve positions. Most of the valve capacity is reached after opening just a small amount.

These three characteristics are shown in Figure 10.9.

When these terms are used to describe a valve they are called intrinsic or inherent flow characteristics. They represent the gain dependence on position in cases where the total dynamic head loss of a system occurs at the valve. As less proportion of system head loss occurs at the valve, the flow characteristic shifts to a relationship called the installed characteristic. This shift is shown in Figure 10.10 for linear and equal percentage valves. Shift in gain is shown in Figure 10.11.

At control valve head losses traditionally used for valve sizing, $DP/DP_D = 0.2$ to 0.33, "linear" valves approach quick open response, and "equal percentage" valves approach linear response. In cases where head loss at the valve is to be reduced to a minimum, the shift in valve response is even greater. When viewing Figures 10.10 and 10.11 keep in mind that the ratio of

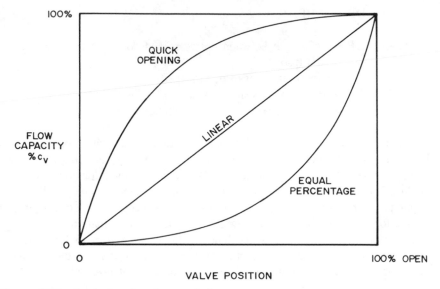

Figure 10.9.   Intrinsic valve characteristics.

valve head loss to total dynamic head loss will not be the same at all flows. This is shown in Figure 10.12. For most of the flow range a valve will be close to its intrinsic characteristic or gain, but near maximum flow the installed characteristic or gain will shift.

Gain analysis is further complicated by the behavior of real valves. Although valves may exhibit a general type of characteristic, the actual installed characteristic will deviate from the ideal. This is especially true at the limits of valve travel. Generally, a valve will provide acceptable control over a portion of its full travel. This range may be expressed as a ratio of high to low flows, high to low valve position, or high to low valve capacity, C_v. The last measure of rangeability is the most useful in selecting a valve. Using the sizing equation, rangeability requirements may be calculated by:

$$\text{Rangeability} = C_{vh}/C_{vl} = (Q_h/Q_l)(DP_l/DP_h)^{1/2}$$

Where: h and l indicate high and low flow rate conditions.

A plot of gain versus valve capacity is shown in Figure 10.13. The valve is a cage-guided, equal percentage globe valve with flow tending to open. It is used to throttle a centrifugal pump. The application called for two-thirds of the total system dynamic pressure losses at maximum flow to occur at the valve. Therefore, the valve should be close to its intrinsic characteristic throughout its operation. Note that the installed characteristic is only equal percentage

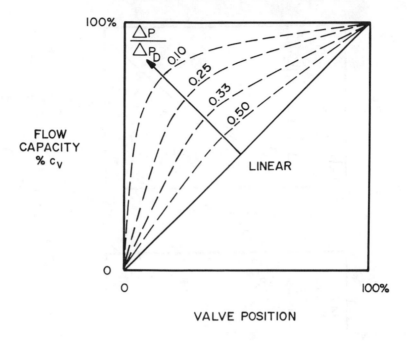

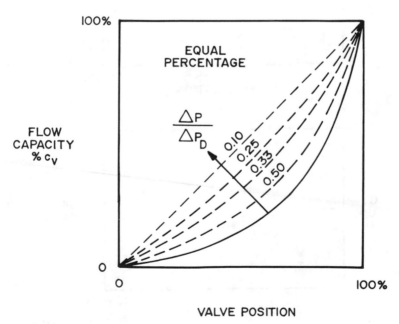

Figure 10.10.   Valve characteristic shift.

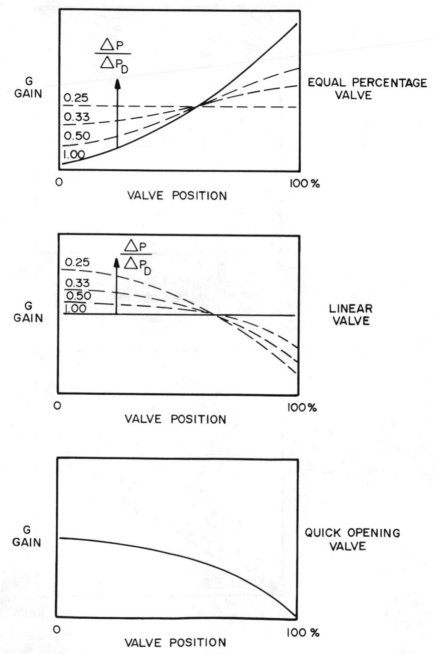

Figure 10.11.   Installed valve gain.

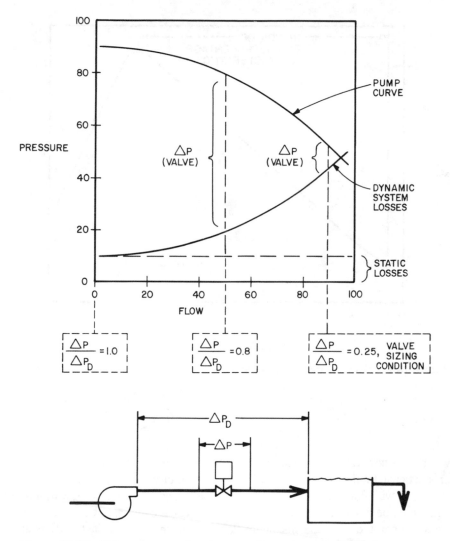

Figure 10.12. Valve pressure drop in a pumping system.

from about 7 to 70% of valve capacity. If the change in characteristic outside this region of equal percentage is detrimental to control of the process. The rangeability is 70:7 or 10:1.

The installed valve gain will affect the control stability. The system should respond to load and setpoint changes with desired dampening and response time. The valve gain should act opposite to the process gain to produce as linear a combination as possible. For example, consider a flow control situation

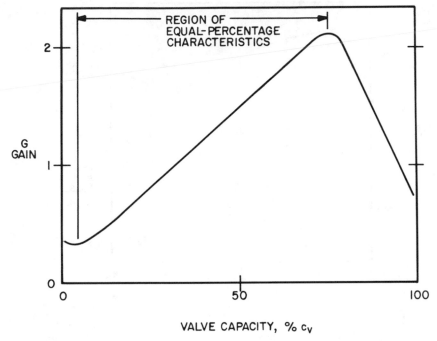

Figure 10.13. Valve characteristic selection guide.

where the flow rate is measured by a differential pressure element. Flow is proportional to the square root of pressure difference. Therefore its gain increases with flow. A quick-opening valve has a gain that decreases with flow and would be a suitable valve choice. A linear valve could also be used if the system operated near the maximum flow, in which case the installed linear valve characteristic would be similar to quick-opening. If the flow transmitter signal was sent to a square root extractor and then to the controller, the meter's gain would be constant. In this case, select a linear valve for operation near the intrinsic characteristic, or select an equal percentage valve for operation near the maximum flow.

Table 10.3 contains a recommendation of valve characteristic for several control applications. If there is doubt about how much installed characteristic to allow for, or which condition overrides, choose equal percentage over linear, and linear over quick-opening.

Table 10.3.   Valve Characteristic Selection Guide

Control Application	Conditions of Application	Valve Characteristic
Flow	Linear flow signal	Linear
	Differential pressure signal	Quick-opening
	Small flow range, large pressure drop changes	Equal percentage
Level	Most applications	Linear
Pressure	Liquid	Equal percentage
	Gas with large pressure drop	Equal percentage
	Gas in fast responding system	Equal percentage
	Gas in slow responding system	Linear

Manufacturers' Options

1. Body styles are available as presented above and in a wide selection of variations. Consult manufacturer's literature for available valve bodies.
2. Materials of body construction:
   a. iron
   b. carbon steel
   c. stainless steel
   d. Hastelloy
3. Materials of trim and plug construction:
   a. carbon steel
   b. stainless steel
   c. brass
   d. copper
   e. Monel
   f. other specialty materials are available
4. Stem packing materials:
   a. TFE
   b. graphite at high temperatures
5. Pipe connections:
   a. threaded
   b. flanged
   c. flangeless (wafer)
6. Body liner material:
   a. Teflon

    b. Buna-N

    c. Viton

7. Trim is available in many configurations for each valve type and varies by manufacturer.

8. Bonnet designs are also varied according to manufacturer.

BIBLIOGRAPHY

1. "Advances in Water and Wastewater Valves," *Consulting Engineer* (September, 1982), p. 105.
2. Boger, H. W., "The Effect of Installed Flow Characteristic on Control-Valve Gain," *ISA Transactions* 8(4): 265 (1969).
3. Ramsey, J. R., and C. D. Fournier, "Matching Valves with Process Requirements." *Instrument and Control Systems* (June 1973), p. 65.
4. Hammit, D., "Choosing a Control Valve is Easy" *Instrument and Control Systems* (December 1976), p. 28.
5. Moore, R. W., "Allocating Pressure Drops to Control Valves," *Instrumentation Tech.* (October 1970), p. 102.
6. Baumann, H. D., "How to Estimate Pressure Drop Across Liquid -Control Valves," *Chem. Eng.* (April 29, 1974), p. 137.
7. Kern, R., "Control Valves in Process Plants," *Chem. Eng.* (April 14, 1975), p. 85.
8. Rezac, M. A., and K. F. Bornhoft, "Choose the Right Control Valve for the Job," *InTech* (October, 1980), p. 59.
9. Vendor literature from: Fisher Controls, Masoneilan International and Pratt Valve, Marshalltown, IA.
10. Driskell, L., *Control Valve Sizing* (Research Triangle Park, NC: Instrument Society of America, 1982).
11. Driskell, L., *Introduction to Control Valves* (Research Triangle Park, NC: Instrument Society of America, 1981).
12. Driskell, L., *Selection of Control Valves* (Research Triangle Park, NC: Instrument Society of America, 1982).
13. Chaflin, S., "Specifying Control Valves," *Chem. Eng.* (October 14, 1974), p. 105.
14. Wolter, D. G., "Control Valve Selection," *Instrumentation Tech.* (October, 1977), p. 55.
15. Wing, P., "Plain Talk on Valve Rangeability," *Instrumentation Tech.* (April, 1978), p. 53.
16. "Flow Equations for Sizing Control Valves," ISA–S75.01–1985, (Research Triangle Park, NC: Instrument Society of America).

Chapter 11

Control Valve Actuators

Control valve actuators are available in a variety of types and sizes. In this chapter, some of the more common actuators are discussed. Included are electro mechanical, pneumatic diaphragms, electric solenoids, pistons, and quarter turn actuators. The terminology used in this chapter follows ISA Standard S75.05–1983, Control Valve Terminology.

ELECTRO MECHANICAL (MOTOR GEAR TRAIN)

The electro mechanical actuator uses an electrically operated motor-driven gear train or screw to position the actuator stem.

Application

Characteristics of motor gear train actuators are:

1. Wide range of speeds and torque outputs available.
2. Adaptable to a wide variety of control circuits and voltages.
3. Compatible with all types of valves and suitable for rising or nonrising stem valves.

4. Means of manual override operation provided.
5. Large units weigh considerably more than their pneumatic and piston counterparts and operate more slowly.
6. Normal fail-safe action is to hold position.

Principle of Operation

Open/Close Operation

A reversing electric motor is used to drive a gear box to open or close the valve. The reversing motor is interlocked to adjustable limit switches to stop the valve motor once the valve reaches a fully open or closed position.

Electrically operated valves generally require high-torque cutout switches in series with the reversing starter to prevent damage to the motor and operating mechanism if the valve becomes obstructed or jammed.

Modulating Operation

A reversing electric motor is controlled by a solid-state reversing motor starter that starts the motor in the required direction when a contact is closed. The motor will continue to operate and change the valve position as long as the contact closure is maintained. When the contact is opened, the motor will stop. A valve position transmitter provides signal conditioning to generate an output signal proportional to the valve position.

Control

Valves can be operated using optional pushbuttons mounted in the actuator or from a remote location. In case of power failure, the actuator can be equipped with a handwheel or similar mechanism to permit manual operation.

For open/close operation, maintained control using either pushbuttons or switches is used. For momentary pushbuttons, a relay is sealed in to move the valve to the full open or full closed position. The valve limit switch breaks the circuit. For maintained switches, the limit switch breaks the circuit.

For modulating operation, control can be either by open/close pushbuttons or by analog position signal. With pushbuttons, the operator must hold the button to move the valve. With a position signal, the operator changes the analog signal value (typically 4–20 mAdc) using a manual loading station or controller in the manual mode.

For modulating operation with pushbuttons or discrete outputs from digital control systems, accurate position control can be obtained for valve with a full range travel time greater than 20 seconds. For valves moving faster than this, a positioner is recommended.

Manufacturers' Options

1. Open/close limit switches:
   a. cam-operated
   b. snap action
2. Auxiliary switches at any point along the stem travel.
3. Position potentiometer typically 0 to 1000 ohms to correspond to 0 to 100% open.
4. Housings:
   a. NEMA 12, dustproof
   b. NEMA 4, weatherproof
   c. NEMA 7, explosion-proof
5. Power supply:
   a. 115 Vac, 60 Hz
   b. 208, 240, 480 Vac, three phase, 60 Hz
   c. 48 Vdc
6. Mechanical brakes to lock a valve in position.
7. A solid-state position controller with an analog (usually 4–20 mAdc) remote setpoint.
8. Housing heaters to prevent condensation, reduce relative humidity, and keep lubricants at proper viscosity.
9. Motor rated for continuous modulating duty.

Design and Installation

- Size actuator for the corresponding valve and service conditions. Actuators are rated by torque which is determined for each application. Most manufacturers have tables that build in safety factors for the various types of valves. These tables allow selection of actuator size directly from valve size or from valve torque data.
- The actuator torque rating must exceed the highest expected valve torque, and the motor must not overheat in the maximum stroke time under the average expected torque. Furthermore, the motor must perform adequately at rated voltage ±10%.
- Frequently, placement of the valve in piping galleries or new process equipment restricts the actuator dimensions and weight. Most often, the smallest and lightest actuator with sufficient torque is selected for the application.

Designer Checklist

Ask the following questions when reviewing or planning a motor gear train actuator installation. If an answer is "no," a motor gear train actuator may not be appropriate.

- Is the valve torque less than 4500 J (40,000 in-lbs)?
- Is a positioner required? Many manual loading stations provide similar functions with contact closure or triac interfaces, so the positioner is not required.
- Is adequate power available?
- Are limit switches or position feedback specified for remote monitoring?
- Is the gear train between the actuator motor and valve stem adequate?
- Is the fail-safe condition to hold the last position? Some spring return electric actuators are available, but review the application to check on merits of electric actuators versus other types of actuators.

Deficiencies

- Modulating service actuators overheat if actuator size is too small.
- Modulating service actuators fail prematurely if operated too frequently.
- Older solid-state triacs for large three-phase motors are not reliable. New triacs have improved reliability.

SOLENOIDS

A solenoid actuator converts electric energy into motion to position the actuator stem or pilot.

Application

Characteristics of a solenoid actuator are as follows:

1. It is used for open/close operation only on pipe sizes of 0.3–8 cm (1/8–3 in.).
2. It cannot actuate large valves or valves with high pressure drops. In this case, the solenoid operates a pilot valve to admit line pressure for operation of the main valve.
3. Fail open or fail close operation is standard. Fail to last position is available for latch-in type solenoids.

4. It is good with short valve strokes.
5. It has fast response speed.

Principle of Operation

A soft iron core moves within the field set up by a coil surrounding the core. Most solenoids use a single sustained level of power energization. To overcome heating problems or to hold last position on power failure, solenoids can have a latch-in plunger action which automatically disconnects the coil from the source.

Control

Manual control is through a pushbutton or switch mounted near the actuator and/or on remote panels. The pushbutton is used for momentary operation and the switch is used for maintained operation. Maintained control may come from device interlocks.

Manufacturers' Options

1. Housings:
    a. NEMA 1, general purpose
    b. NEMA 3R, rainproof
    c. NEMA 4, weatherproof
    d. NEMA 6, submersible
    e. NEMA 7, explosion-proof
2. Power supply:
    a. 24, 115, 220, or 480 Vac, 60 Hz
    b. 6, 12, 24, 120, or 240 Vdc
3. Operation:
    a. normally closed (open when energized)
    b. normally open (close when energized)
    c. latch-in
4. Manual Operation:
    a. push type—spring return for momentary operation
    b. screw type—manual return for maintained operation

Design and Installation

- Size the actuator for the corresponding line size and service conditions.
- Specify a housing suitable for the environment.

Designer Checklist

Ask the following questions when reviewing or planning a solenoid actuator installation. If an answer is "no," a solenoid actuator may not be appropriate.

- Is the line size less than 8 cm (3 in.)?
- Can the solenoid handle the maximum differential pressure across the valve?
- Is the minimum line pressure sufficient to keep the solenoid open?
- Is the coil class available for the temperature range expected?

Deficiencies

- It is very difficult to get a positive indication of valve status. Most solenoid-actuated valves do not have a limit switch option.

PISTONS

A piston is a fluid-powered device in which the fluid acts upon a movable cylindrical member, the piston, to provide linear motion to the actuator stem. A piston may be either pneumatic-type using air, or hydraulic-type using oil or water.

Application

Characteristics of piston actuators are as follows:

1. High-pressure air or hydraulic fluid is generally supplied from a common pumping unit.
2. Actuator movement is accomplished through a solenoid-operated pilot valve.
3. Piston actuators are capable of high-torque outputs.
4. They are capable of long stroke outputs.
5. They have a normal fail-position action of hold, fail open or fail close.
6. They are available in single or double acting configurations.

Principle of Operation

Single Acting

A three-way solenoid pilot valve directs air or hydraulic fluid to one side of the piston. A spring on the other side of the piston opposes the force of the fluid. The fluid causes the spring to compress. Upon release of the pressure, the piston moves back to its "at rest" position.

Single acting actuators are used in applications requiring the valve to move to a fail-safe position on loss of pressure. Figure 11.1 shows a single acting piston actuator.

Double Acting

A four-way solenoid pilot valve directs air or hydraulic fluid to one side of the piston while relieving pressure on the other side.

Double acting cylinders are generally used in applications requiring the valve to hold last position on loss of pressure.

Piston-actuated modulating valves are typically butterfly, ball, and plug type valves. Pistons can be used for sluice gates and movable weirs.

Adjustable limit switches detect the fully open or closed position of the valve. A slide wire position potentiometer provides a resistance output signal proportional to the actuator stem position for remote monitoring.

Control

Manual operation at the valve is by operating the solenoids, either electrically or manually. In the case of pressure loss, the valve is equipped with a handwheel or similar mechanism to permit manual operation.

For open/close operation, maintained control using either pushbuttons or switches is used. For momentary pushbuttons, a relay is sealed in to actuate the appropriate pilot solenoid and move the valve to the full open or full closed position.

For modulating operation, control can be either by open/close pushbuttons or positioner. With pushbuttons, the operator must hold the button to move the valve. With a positioner, the operator must change an analog signal value (typically 4–20 mAdc).

For modulating operation with pushbuttons or discrete outputs from digital control systems, accurate position control can be obtained for valve with a full range travel time greater than 20 seconds. For valves moving faster than this, a positioner is recommended.

Manufacturers' Options
1. Open/close limit switches
2. Positioner:
    a. force balance
    b. electrohydraulic servo systems
3. Slide wire potentiometer

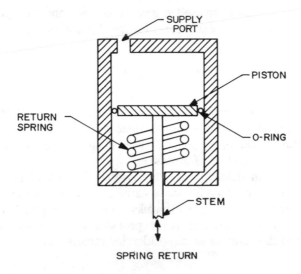

Figure 11.1.   Piston actuator.

Design and Installation

- Size the actuator for the corresponding valve and service conditions. Pistons are rated by force.
- Minimize fluid piping lengths.
- Size the pneumatic or hydraulic system to handle the piston displacement and the number of actuators being operated from the same source.
- For isolated valves or to provide backup, a local pump powered by an electric motor, compressed gas, or the process liquid can be used to provide the pressurized fluid. A small pump can supply high pressure oil or water to the piston. Limit sensing devices should be used to stop the motor or to operate valves that shut off the fluid line.
- For large sluice gates, it is possible to use pistons with the actuator stem from both sides of the piston. One stem is connected to the gate while the other end is connected to a counterweight. This significantly reduces the piston size and pressure requirements.

Designer Checklist

Ask the following questions when reviewing or planning a piston actuator installation. If an answer is "no," a piston actuator may not be appropriate.

- Does the fluid system provide the desired force and speed of operation?
- For large valves and gates, is there clearance for the actuator?
- Is a positioner required?
- Are limit switches or position feedback specified for remote monitoring?
- Does the actuator type, single or double acting, match the desired fail-safe position of the valve?
- Will ambient temperature changes affect force required to move valve? Is actuator sized to handle this?

Deficiencies

- Leakage of the piston seals, fluid lines causes loss of force or drift if no positioner is provided.
- The valve becomes stiffer with temperature or age. Piston not sized to handle this condition.
- Fluid pressurizing system cannot handle load placed by multiple actuators acting at the same time.
- Wear on pressuring system causes reduced pressure.
- Sloppy linkage causes large deadband before movement occurs.

PNEUMATIC DIAPHRAGM

The energy of a compressible fluid, usually air, acts upon a flexible member, the diaphragm, to provide linear motion of the actuator stem.

Application

Pneumatic diaphragm actuators react to external pressure sources which may be the output pressure of a controller, an electric to pneumatic converter, a manual loader, or a positioner.

Diaphragm actuators are available in two types, the spring and diaphragm and the springless diaphragm. A spring and diaphragm actuator is shown in Figure 11.2.

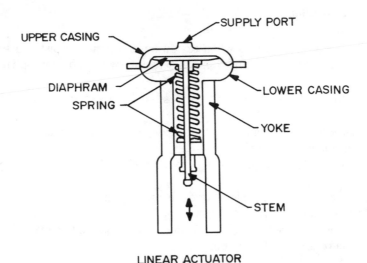

LINEAR ACTUATOR

Figure 11.2.   Pneumatic diaphragm actuator.

Characteristics of a pneumatic actuator are as follows:

1. The pneumatic actuator offers high reliability at low cost and is easy to maintain.
2. It is fast acting.
3. It is available for either fail-open or fail-closed action.
4. It normally uses air signal ranges of 3–15 psig.
5. It has limited thrust capability. Normally limited to valves 8 in. and smaller.

Principle of Operation

Control valves usually have relatively long stem travel and stem position should be linearly proportional to the instrument output pressure (usually 3–15 psig) on the diaphragm.

For a spring and diaphragm actuator, pressure on the diaphragm causes the stem to move. The force equals the pressure times the diaphragm area. This force is opposed by a spring. The valve and spring design permits a linear relationship between instrument air pressure and stem position. This assumes that the process flow force on the valve does not change over time.

The disadvantage of a spring is that its force is a constant at any one position. It cannot overcome varying friction and differential pressure forces. In a springless diaphragm actuator, the pressure on both sides of the diaphragm varies so that the sum of the two always equals the supply pressure. With a positioner, any reasonable supply pressure can be used.

Pneumatic diaphragm-actuated modulating valves are typically butterfly, ball, and plug type valves or bladder type pinch valves.

Control

Remote position control from an external source is usually from pneumatic instruments. Current-to-pneumatic converters can be used for electrically based control systems.

Manufacturers' Options

1. Pressure ranges:
   a. 6–30 psig
   b. 12–60 psig
   c. 3–27 psig
2. Positioners:
   a. position balance
   b. force balance
   c. electropneumatic

Design and Installation

- Size the actuator for the corresponding valve and service conditions. Actuators are rated by force.
- Minimize pneumatic piping lengths.

Designer Checklist

Ask the following questions when reviewing or planning a pneumatic actuator installation. If an answer is "no," a pneumatic actuator may not be appropriate.

- Is the stem travel sufficient for the valve?
- Does the actuator match the instrument output pressure?
- Can the actuator handle the maximum differential pressure expected?
- Can the actuator seat and unseat the valve?
- Have you accounted for the change in effective diaphragm area? The applied loading pressure to close a valve must be higher than the value given by calculations which assume a constant area. Some actuators are designed to present a constant area.
- Is a positioner required?
- Are limit switches or position feedback specified for remote monitoring?

- Will ambient temperature changes affect force required to move valve? Is actuator sized to handle this?

Deficiencies

- Pneumatic actuator is difficult to position accurately without a positioner.
- Leaks in pneumatic system slow response.

QUARTER TURN

Quarter turn actuators are not a class of actuators. However, their use is common enough to include here.

Motor gear train, piston, and pneumatic actuators can have a problem stopping a valve in the exact open and closed position. Limit switches can be used to operate the motor or pilot solenoids but it is difficult to set and keep limit switches in adjustment.

Application

Characteristics of quarter turn actuators are as follows:

1. Either piston or diaphragm actuators can be used.
2. Use where full closing of a valve to avoid leakage is important.
3. Appropriate for butterfly, ball, and plug valves.

Principle of Operation

Two types of quarter turn actuators are available: the rack and pinion type and the linkage type.

For the rack and pinion type, a piston or diaphragm actuator moves a rack back and forth past a pinion. The linear motion of the piston is converted to circular motion. Precision adjusting screws are located at both ends of the cylinder to stop the piston at the desired position. There is no reliance on limit switches to stop the valve in the correct position.

A double acting quarter turn piston actuator is shown in Figure 11.3.

For the linkage type actuator, the output force of the piston or diaphragm actuator is obtained by transferring the force through an arm to convert it to torque.

Figure 11.4 shows a double-acting piston linkage type actuator and Figure 11.5 shows a diaphragm-actuated linkage type actuator.

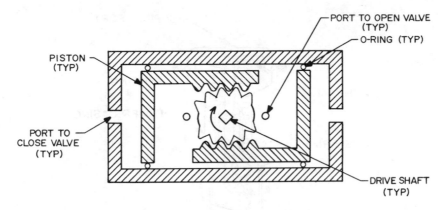

DOUBLE ACTING, RACK & PINION

Figure 11.3.   Double acting piston, rack and pinion actuator.

Control

Both open/close and modulating operation is available for these actuators. Controls would be the same as that for the piston and diaphragm actuators discussed previously.

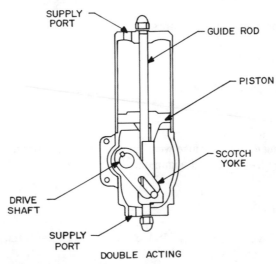

DOUBLE ACTING

Figure 11.4.   Double acting piston, linkage actuator.

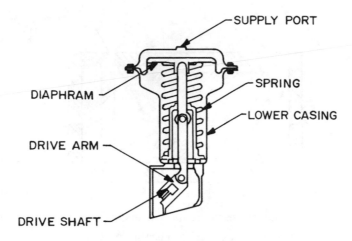

ROTARY OR QUARTER-TURN ACTUATOR

Figure 11.5. Diaphragm, linkage actuator.

Manufacturers' Options

Options are similar to those for piston and diaphragm actuators. For position monitoring, either potentiometers or slide wires can be used because there is both rotary and linear motion.

Design and Installation

See piston or pneumatic actuators.

Designer Checklist

See piston or pneumatic actuators.

Deficiencies

- Modulating duty may cause wear on rack and pinion or linkage if operation is too frequent or if alignment is improper.

BIBLIOGRAPHY

1. "Control Valve Terminology," ISA–S75.05–1983. (Research Triangle Park, NC: Instrument Society of America).
2. Beard, C.S., *Final Control Elements* (Philadelphia, PA: Chilton Co., 1969).

Chapter 12

Process Control Instrumentation

Process control instrumentation includes pushbuttons and switches, recorders and annunciators, controllers and computing modules, relays and timers, drum programmers, programmable logic controllers, computers, and a multitude of similar devices. As such, a detailed discussion is beyond the scope of this chapter.

The first part of this chapter describes control strategies applicable to various processes. Following the control strategies is a brief discussion of operational considerations and process control system options. Finally, implementation considerations are discussed.

The purpose of process control instrumentation is to improve the operation of the process. The instrumentation required to do this depends on the size of the plant, the complexity of the processes, the variability of the influent, and effluent quality requirements.

To select process control instrumentation you should perform the following steps:

1. Define the control strategy for each process.
2. Define the monitoring needs for each process.
3. Define the failure operations for each process.

4. Identify resource constraints.
5. Identify physical constraints.
6. Select the process control instrumentation system that best meets the above criteria.

CONTROL STRATEGIES

Any undesired change to a process input is a disturbance. Disturbances can include changes in influent quality, flow or pressure changes, and equipment in-service status. Control is any action taken to maintain a process at the desired operating point. Control actions include two broad categories, discrete and analog (continuous). Discrete control is sometimes referred to as digital control but the term is not used here in order to avoid confusion with computer control.

Control actions may be implemented manually or automatically. The choice of manual or automatic control is dependent on how often a correction is needed, the ability of the process to tolerate disturbances, the magnitude of the disturbances, and the safety of personnel or equipment.

Discrete Control

Discrete control is characterized by on/off or open/close type control actions. These actions are in response to a predefined program of events, elapsed time, or an analog value reaching some preset limit. Discrete control includes simple on/off control of a single device such as a sump pump, sequential startup/shutdown of a complex device such as an aeration blower or centrifuge, and a sequence of operations such as a filter backwash operation. Table 12.1 lists some discrete control applications.

Analog (Continuous) Control

Analog control is characterized by variables which can be continuously observed and represented. For most water and wastewater plants variables are represented by 4–20 mAdc, 1–5 Vdc, or 3–15 psig signals.

Open-Loop Control

A generalized block diagram of an open-loop control system is shown in Figure 12.1. The final control element (modulating valve, variable speed pump, etc.) is set at one point within its operating range. The distinguishing feature of open-loop control is that a continuous measurement of the variable to be controlled is not available or is not connected to the controls. There is no

Table 12.1. Discrete Control Applications

Process	Controlling Variable
Sumps	Level switch
Filter backwash	Filter head loss Filter level Time since last backwash Time of day
Primary sludge pumping	On-time/off-time
Intake screens	Repeat cycle timer Head loss Accumulated flow
Digester charging and withdrawal	On-time/off-time
Centrifuge startup/shutdown	Automatic request Manual request
Plate and frame press	Cycle timer
Centrifugal blowers	Pressure trip

assurance that the control objective is actually being achieved. Control will be satisfactory only if the final control element is properly set and no disturbances occur. If the original conditions are disturbed, operator intervention will be required to maintain balance within the process. Open-loop control is

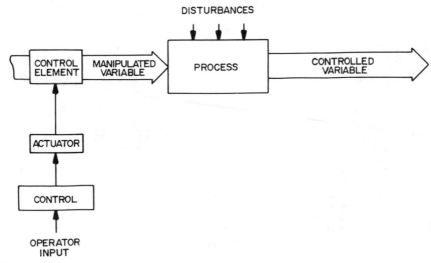

Figure 12.1. Open-loop control.

satisfactory for applications where conditions do not vary significantly and close regulation of the process is not required.

Closed-Loop Control

Figure 12.2 is a block diagram of a closed-loop control system. In this case, a continuous measurement of the controlled variable is made and routed to the controller. The controller compares this measurement or feedback signal with the set point (the desired value). If there is any difference between desired and actual, the controller outputs a corrective action to the final control element. The term "closed-loop" describes the path that is formed around the process by the controlled variable measurement, the controller, and the output to the final control element. The effects of any change in the system propagate around the loop.

The flow control system shown schematically in Figure 12.3 is an example of closed-loop control. The objective is to maintain the flow rate at some desired value by adjustment of a modulating valve. In this example, flow is the controlled variable. The valve position is manipulated by the controller to eliminate any deviation of the flow from the set point value. The valve is the final control element and its position is referred to as the manipulated variable.

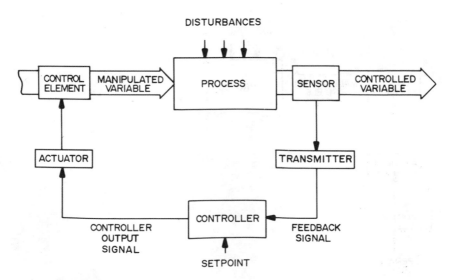

Figure 12.2. Closed-loop control.

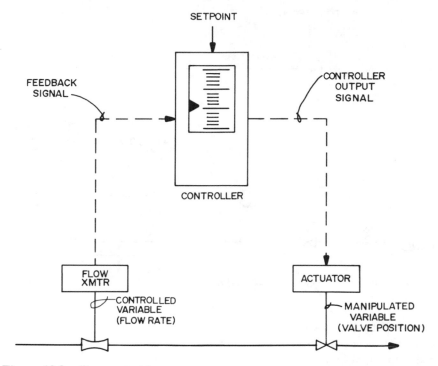

Figure 12.3. Flow control loop.

Cascade Control

For applications where the time to respond to process disturbances is long, improved control can be obtained through the use of cascade control. In this type of control, two complete closed loops are formed, one within the other.

To illustrate cascade control, consider the dissolved oxygen (DO) control system shown in Figure 12.4. The variation in DO is used to manipulate the position of the air valve. Due to the large process capacity, disturbances in the air flow are not immediately sensed by the DO analyzer. When the DO measurement begins to change, the control system will begin to modify the air flow to the aeration basin. There may be so much time lag that the control is ineffective or unacceptable. For example, by the time the effect of a disturbance in the air supply pressure is sensed by a change in DO, the pressure disturbance may have subsided or the pressure may have changed again. The correction signal from the controller based on DO may be inappropriate for the current air supply conditions.

There are two sources of disturbances to air supply pressure. In a multiple tank system where all of the tanks utilize a common air supply, the changing

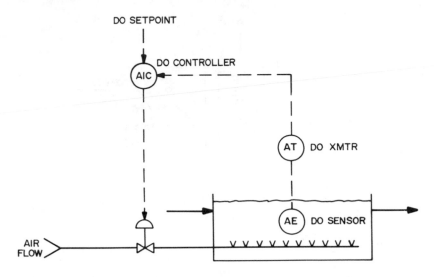

Figure 12.4. Direct control of dissolved oxygen.

demands of each tank can disturb the air supply pressure. Secondly, where multiple blowers are used to match the demands of the process, the starting and stopping of blowers introduces pressure disturbances.

Figure 12.5 illustrates the addition of a control loop for air flow with its set point derived from the DO controller. The DO controller is said to be "cascaded" into the air flow controller.

The DO control loop is referred to as the outer loop. The air flow control loop is the inner loop. With this arrangement, disturbances in the air supply pressure that would affect the flow are now corrected by the inner flow control loop. Corrective action is now initiated without having to wait for the DO to change, as was the case under single loop control.

Feedforward Control

Feedback control may not be satisfactory in some processes for three reasons. First, the control system does not act until after a disturbance has caused an error. Second, the effect of the correction is not sensed until after it has propagated around the entire control loop. This is not a problem in fast-responding control loops such as flow control. It can be a problem in loops with long time periods between making corrections and sensing the effect of the correction, such as chlorine residual control. Third, it may not be possible to measure the controlled variable as in the case of flocculation or clarifier settling efficiencies.

Feedforward control can be used as an alternative to feedback control. It

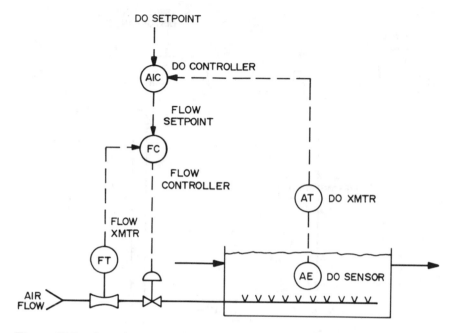

Figure 12.5.   Cascade control of dissolved oxygen.

predicts how much corrective action will be required because of a disturbance. Feedforward control measures one or more inputs to a process. Any change in the inputs causes a corresponding change in the manipulated variable (see Figure 12.6). If the exact amount of corrective action required can be predicted, no deviation of the controlled variable will occur. In practice, this is difficult to achieve because all of the disturbances cannot be accounted for. Feedforward control is a form of open-loop control because the controlled variable is not used in the loop.

The chlorination process is dominated by dead time, the time between the chlorine addition change and the sensing of the effect of the change. The dead time is due to the time to pass through the chlorine contact chamber and the time to transport the sample the chlorine residual analyzer. The primary disturbance to which this system is subjected is the change in demand for chlorine. Using a feedforward approach, the rate of chlorine application can be based on a measurement of the inflow as shown in Figure 12.7. If the characteristics of the influent do not vary, this strategy can provide good control of the chlorine residual. This strategy works well on finished water and tertiary treatment plant effluent. For most secondary effluents, changes in solids concentration make this control unsuitable.

The above control is a special case of feedforward control referred to as ratio

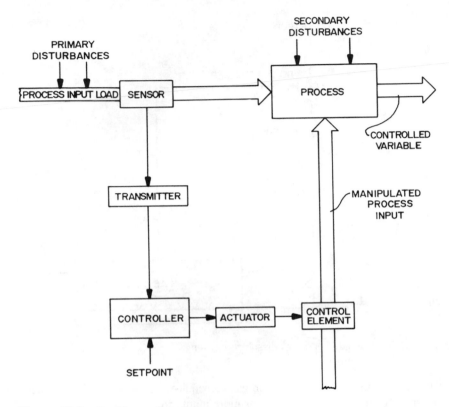

Figure 12.6.   Feedforward control.

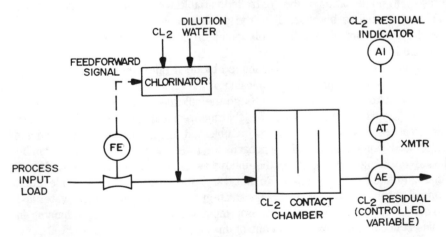

Figure 12.7.   Feedforward control of chlorination.

control or flow-pacing control. It is commonly used in processes involving chemical feed because there is no practical way to measure the controlled variable in order to apply feedback control.

Feedforward control is not adequate if strict regulation is required. In these cases it is necessary to add feedback to obtain the advantages of both types of control. Feedforward control provides advance compensation for major disturbances sensed in the process inputs. Feedback control provides a trimming effect to correct for minor disturbances.

Table 12.2 lists some analog control applications.

Adaptive Control

Adaptive control occurs when the control system automatically corrects for changes in the process behavior. Examples include self-tuning controllers and model-reference control. Self-tuning controllers constantly monitor the loop performance and adjust the controller tunings to provide the desired system response. In plants with many control loops, maintaining good tuning can be a time-consuming activity.

Loops with significant dead time or loops subject to disturbances which are best regulated using feedforward action are candidates for adaptive control.

Combined Control

Many water and wastewater control systems combine discrete and continuous control. These controls enable security, safety, and optimum performance. Startup and shutdown logic, failure detection, and process characteristic tracking can be used to minimize dependence on operator-directed corrective actions and maintain performance. Table 12.3 lists some examples of combined control applications.

Expert Systems

Expert systems technology is a branch of artificial intelligence that uses computers to match or exceed the decisionmaking capability of human experts. Expert systems are useful to solve problems where no mathematical or algorithmic solution exists because data is inexact, uncertain, or based on probabilities.

Expert systems use groups of rules to perform consistent control actions. The system can monitor the process continuously to diagnose alarm conditions and prompt the operator on how to deal with them. An expert system can automatically implement corrective action.

Expert system applications include combined sewer overflow control and storm water routing, secondary treatment operational options, and filter control.

Table 12.2. Analog (Continuous) Control Applications

Process	Controlled Variable
Feed Forward	
Chlorination (water, tertiary effluent	Flow pace chlorinator
Flocculation (alum, ferric sulfate)	Flow pace chemical pump
Water softening (lime, soda ash)	Flow pace chemical pump
Settling aids (polymer)	Flow pace chemical pump
Feedback (closed loop)	
pH	Proportional/integral
Digester temperature	On/off gap control
Hearth temperature	Proportional/integral/derivative
RAS to aeration tank	Proportional/integral
Vacuum filter vat level	Proportional/integral
Centrifuge back drive	Proportional/integral
Filter effluent flow	Proprotional/integral
Cascade	
Flow splitting	Outer loop: pressure, level, master valve position, most open valve position. Inner loop: flow
Dissolved oxygen	Outer loop: dissolved oxygen Inner loop; air flow
Centrifuge polymer	Outer loop: centrate solids Inner loop: polymer flow
Blower header pressure	Outer loop: pressure Inner loop: blower current/power
Feedforward/Feedback	
Chlorination (secondary effluent)	Feedforward: percent of influent Feedback: chlorine residual
Clarifier sludge withdrawal	Feedforward: percent of influent Feedback: solids density or flow
Chemical feed (tank/valve systems)	Feedforward: percent of influent or solids loading Feedback: chemical flow
Adaptive	
Total waste sludge	Daily mass or 24 hours continuous depending on sludge solids changes
Incineration	Autogenous combustion with temperature profiling
Vacuum filter	Maximize sludge cake dryness with chemical addition and drum speed depending on sludge characteristics

Table 12.3. Combined Control Applications

Process	Controlled Variables
Booster station	Pump on/off, pressure
Lift station	Pump on/off, level
Blower header	Blower on/off, pressure
pressure	to current cascade

Expert systems are appropriate when:

1. The current methods are too slow, inaccurate, or inconsistent.
2. The problem is small enough that it can be solved by a human expert in a matter of hours, not days, weeks, or months.
3. The human expert exists. For new plant construction, no operational experience is available and expert systems are not appropriate.
4. Only one or two experts exist when many are needed. This is particularly appropriate to shift work.
5. There is an identifiable benefit.

OPERATIONAL CONSIDERATIONS

Monitoring Needs

Once the desired control for each process is defined, monitoring needs can be established. Monitoring includes:

- Alarm detection and annunciation.
- Equipment status and run time.
- Trending.
- Flow totalization.
- Performance calculations.
- Operational reporting.
- Historical data acquisition and storage.

Control Philosophy and Failure Operations

The control philosophy affects the instrumentation system selection. Control philosophy considerations include the following:

1. Manual or automatic controls.
2. Momentary or maintained pushbuttons and switches.
3. Analog (4–20 mAdc) or incremental (triac) control for continuous control loops.
4. Controls located local near device or remote from equipment.

5. Each loop controlled by individual devices, several loops controlled by one device, many loops controlled by one device.
6. Type of backup control required for each loop.

Resource Constraints

Resource constraints which will affect the selection of the process control instrumentation include:

1. Operation and maintenance personnel skills.
2. Time and budget limitations.
3. Geographical location of plant.
4. Political/social issues such as using local suppliers, pay scales, union agreements, community acceptance, etc.

Physical Constraints

Physical constraints which will affect the selection of the process control instrumentation include:

1. Environment in which the equipment must operate including:
   a. temperature and humidity
   b. airborne contaminants such as dust, corrosive gases, and explosive gases
   c. power reliability and quality
   d. lightening, earthquakes, etc.
2. Distances between equipment and buildings, plant layout, physical and process control areas within the plant.

PROCESS CONTROL SYSTEM OPTIONS

There is a great variety of process control instrumentation available today. In the past, we could think in terms of individual panel-mounted devices and/or a computer control system. Today there are a variety of "computer" control systems and no single approach is the best for every application. Rather, each application must be evaluated and appropriate selection made.

Process control automation has evolved along with digital technology. In the early 1960s, analog control on a loop-by-loop basis was the accepted approach. Any optimization was handled by the operator. In the early application of the computer, analog control was replaced with the digital counterpart. While this was usually of great economic benefit, it soon became apparent that the most far-reaching advantage of the computer was that it could be programmed to perform massive calculations and to make decisions rapidly and consistently.

In the 1970s, utilization of computers in many industries increased greatly with the advent of the minicomputer. This technology offered an economical alternative to manual and analog control systems even for small plants or installations. With the continued evolution of technology into the large-scale integrated circuits, it became economically possible to distribute the computers around the plant. Today microcomputers are located in "analog" controllers and in individual instruments.

Because of the variety of instrumentation and because of technology changes, it is difficult to classify instrumentation by technology used. Almost all process control instruments use microprocessor-based technology. This includes timers, panel mounted controllers, and similar equipment. Therefore, process control instrumentation is classified by function.

The functional categories of process control instrumentation systems include the following:

1. Self-contained, single loop.
2. Shared, multi-loop.
3. Supervisory set point.
4. Distributed, networked.

Self-contained, single-loop systems consist of individual relays, timers, controllers, etc. Shared multi-loop systems consist of multi-loop controllers and computers which operate and display information independent of one another. Supervisory set point systems have a computer which generates set points or other control information for use by both self-contained and shared control devices. Networked systems are characterized by data highways and shared operator displays. Networked systems can include micro-processor based instruments, single- and multiple-loop controllers, process controllers, programmable logic controllers, operator display stations, personal computers, minicomputers, and large computers.

Self-Contained, Single Loop Control Systems

Self-contained, single loop control systems consist of instruments mounted in control panels and include:

1. Pneumatic instruments.
2. Electronic analog signal instruments such as adders, subtractors, and three-mode controllers. These instruments may be transistor or microprocessor based.
3. Electronic digital signal instruments such as sequencers, timers, lights, and relay replacers. Both transistor-based and micropro-cessor-based instruments are available.

4. Mechanical instruments such as relays, switches, drum programmers, and timers.

Control panels with their collection of switches, relays, indicator lights, analog indicators, totalizers, recorders, annunciators, manual loading stations, computing functions, and three-mode controllers have been the traditional process control instrumentation for more than half a century.

Field instruments are connected to the unit process or area control panels. The grouping of controls make possible interrelated control functions. Discrete information such as pump running, overloads, or critical alarms are shown on the panel with lights and annunciators. Control functions on a control panel are pushbuttons and selector switches. Automatic sequencing may or may not be included.

Small local control panels are provided with valve actuators, motors, and process equipment. They can be used for operation but are generally intended for maintenance. Larger panels are custom designed to operate a process or plant area. An operator can view the operation and make control adjustments through these panels. The panels may include programmable logic controllers as relay replacers for equipment sequencing.

Although good for small plants or as a means of backup control, panels and stand-alone systems are weak in the following areas:

1. Plant-wide operating information is not consolidated.
2. Processes which are in series are operated independently. Required interprocess operator communication and control is usually verbal via telephone. Some key process variables could be wired to multiple panels in order to minimize communication errors.

Shared, Multi-Loop Control Systems

In a shared time system, one controller divides its computation or control time among several loops. Examples of shared, multi-loop controllers include the following:

1. Microprocessor-based, panel-mounted multiple loop controllers.
2. Programmable logic controllers (PLCs).
3. Distributed process controllers (DPCs).
4. Personal computers and minicomputers.

Supervisory Set Point Control Systems

This method of control combines distributed single loop or multi-loop controllers with the data reporting and calculating capabilities of the computer. In supervisory systems, all control is accomplished by local controllers, including analog (continuous) and digital (discrete) sequential control. Control decisions and adjustments are initiated by the computer system.

Process inputs are transmitted to the panel-mounted controllers and then to the computer. The computer will log and retain the information. It will prepare all reports and annunciate alarm conditions as they appear.

The computer will also contain programs which will analyze the process inputs and will decide when certain process changes or adjustments are necessary. Typically, only loops which can be benefited by optimization will be included. Computer control outputs are sent to the panel-mounted controllers in the form of set point changes and sequence logic initiation. This type of control system is not common in water and wastewater treatment processes.

Distributed, Networked Control Systems

The major characteristic of a distributed, networked system is the transmission of control and operating information to and from user-defined locations over a data highway. In contrast, a shared control system may contain similar components but the components operate as stand-alone devices. For example, a programmable logic controller (PLC) may be used to control several loops and an operator display may be connected to the PLC. This is a shared, multi-loop control system. If two PLCs and the operator station are connected by a data highway, then the system is a distributed control system.

Distributed control systems can range from very small, containing 3 or 4 control loops and 50 to 100 inputs and outputs, to very large containing 10,000 or more inputs and outputs and several hundred control loops. Examples of distributed control systems include the following:

1. Two or more single or multi-loop controllers and a personal computer.
2. Programmable logic controllers (PLCs) and a personal computer.
3. PLCs with dedicated operator display stations and personal computer data logger.
4. Distributed process controllers (DPCs) with an shared display station (operator station).
5. DPCs or PLCs with operator stations and gateway to a minicomputer.

There are many different configurations for networked systems which can be used in water and wastewater plants. Figure 12.8 combines a number of different components into a distributed system for example purposes. It is unlikely that a system would contain all of the components shown.

In Figure 12.8, the dashed box with the number 1 in the upper left corner contains a single-loop controller which can be self-contained or can operate as part of a distributed control system. Area 2 shows how single- and multi-loop controllers can be integrated into a distributed control system with a local

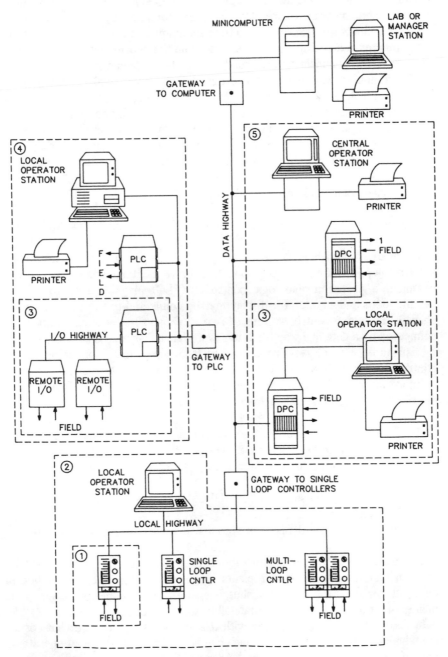

Figure 12.8.   Distributed control configuration.

highway and an operator display station. The two areas labeled 3 show how a PLC or DPC can be used as a shared, multi-loop control system without integration into a distributed network. Area 4 shows how PLCs can be integrated into a distributed network. Area 5 shows DPCs in a network with a central operator station.

Some other features of distributed control systems are as follows:

1. The central minicomputer is not needed for control or alarm logging. Most operator stations can be furnished with an alarm printer. Some stations can even print operating reports.
2. A personal computer or personal computer network could be used instead of the minicomputer.
3. PLC-based systems can have their inputs and outputs located remote from the PLC. This feature can save wiring costs and can allow placement of the input/output equipment in more severe environments.
4. The squares labeled "gateway" are devices to connect different types of equipment to the data highway. The gateways convert the data highway protocol to local equipment protocols so that dissimilar devices may communicate data and control commands.

In the manufacturing sector, General Motors and others have defined a standard protocol, the Manufacturing Automation Protocol (MAP). This was needed to provide integration of the large number of PLCs in a typical assembly plant. In water and wastewater applications, requiring MAP adds additional complexity to the control system and is of questionable value. There is some concern that MAP may not be suitable in a process control environment.

As is evident, an advantage of distributed control systems is they can be assembled in a building block fashion. You can start small and add process control instrumentation as budget and needs dictate.

PROCESS CONTROL SYSTEM IMPLEMENTATION

Applications

Table 12.4 lists applications for the various process control instrumentation systems discussed. The control function is the dominant control. Most plants will fall into the combined control applications with some backup control functions from panels using other methods shown.

The preceding applications are also affected by the monitoring functions, resource constraints, physical constraints, and reliability/redundancy requirements for failure operations.

Table 12.4. Control System Applications

Control Function	Process Control System
Discrete	Mechanical relays Programmable logic controllers
Open loop	Self-contained, single loop controllers Pushbuttons and switches
Closed loop, analog	Self-contained, single loop controllers Multi-loop controllers Distributed process controllers
Cascade	Multi-loop controllers Distributed process controllers
Feedforward	Self-contained, single loop controllers Multi-loop controllers Distributed process controllers
Adaptive	Distributed process controllers Supervisory set point with self-tuning single loop controllers Computers
Combined	Multi-loop controllers Programmable logic controllers Distributed process controllers Computers

Simple monitoring requirements can be performed by panel-mounted instrumentation such as alarm annunciators, lights, and totalizers. For some systems, personal computers or operator stations can record alarms, perform calculations, and print reports. For large plants with extensive reporting and data analysis needs, personal computer networks and minicomputers can be used.

Resource constraints can affect the equipment manufacturers allowed to provide the control system. In general, programmable logic controllers and personal computers are available locally and are easier to maintain.

PLCs can tolerate a more severe environment than distributed controllers. Failure operations affect the design of panels and controller outputs as well as valve actuators and pump controls.

Installation

- Avoid running 4–20 mAdc signals more than 600 m (2000 ft) without boosters.
- Allocate one-half day or more per loop for checkout. Check the loop from the field device through panels, controllers, operator stations, and computers.

- Use separate conduits for analog and low level digital control signals.
- Ensure that the environment is suitable for the equipment. Provide air conditioning and air scrubbing to maintain the contaminant levels at or below those shown in Table 12.5.
- Avoid running data highways more than 1500 m (5000 ft).
- Use fiber optic data highways between buildings in areas subject to severe lightning or in areas of high electrical noise.
- Provide clean power sources for microprocessor-based equipment.
- Provide false floors for minicomputer rooms.

Designer Checklist

If you can answer "yes" to the following questions, the process control instrumentation should be correct for the application.

- Will the user be able to support and maintain the system? If not, is local outside support available?
- Is the level of backup control correct for the process application?
- Does the control system complexity match the process control needs?
- Have you considered the human engineering in control panel layout? Have you grouped devices according to the process layout or associated with process flow?
- Is momentary or maintained control used consistently? Are switches labeled consistently?
- If 4–20 mAdc control is used, do valve actuators and pump speed controllers move to or stay at a safe position on loss of signal?
- Have you considered the human engineering in graphic display layout? Show information according to the process layout or associated with process flow. Use a top-down (overview to detail) approach to avoid crowding too much information on one display.
- Have you determined who has system responsibility?
- Have you verified that equipment interfaces are proper?
- Have you required factory and field testing?

Table 12.5. Maximum Containment Levels

Contaminant	Maximum Level		
	Mild	Moderate	Units
Dust	1000	5000	micrograms/m^3
H_2S	3	10	parts/billion
Cl_2	1	2	parts/billion
SO_2, SO_3	10	100	parts/billion
NO_x	50	125	parts/billion
Copper Reactivity Level	300	1000	angstroms

- Is the operation and maintenance training adequate?
- Have you considered future equipment compatibility for distributed control systems?

Maintenance

1. Do you have a maintenance contract?
2. Are spare parts readily available now? In the future?
3. Will maintenance be board replacement or component repair? This will affect the level of training, on-site spare parts complement and type of test equipment required.
4. Who will "maintain" the software? Many single-loop controllers are programmable.

BIBLIOGRAPHY

1. "Automation and Instrumentation", AWWA Manual M2 (Denver, CO: American Water Works Association, 1983).
2. Glysson, E. A., et al., *Computerization in the Water and Wastewater Fields* (Chelsea, MI: Lewis Publishers, Inc., 1986).
3. "A Primer for Computerized Wastewater Applications," Manual of Practice No. SM-5, (Alexandria, VA: Water Pollution Control Federation, 1986).
4. "Process Instrumentation Terminology," ANSI/ISA-S51.1-1979 (Research Triangle Park, NC: Instrument Society of America, 1979).
5. "Graphic Symbols for Distributed Control/Shared Display Instrumentation, Logic and Computer Systems," ISA-S51.3-1983 (Research Triangle Park, NC: Instrument Society of America, 1979).
6. "Instrumentation Symbols and Identification," ISA-S5.1-1983 (Formerly ANSI Y32.20) (Research Triangle Park, NC: Instrument Society of America, 1979).
7. "Functional Diagramming of Instrument and Control Systems," SAMA Standard PMC 22.1-1981 (Washington, DC: Scientific Apparatus Makers Association, 1981).
8. Considine, D. M., *Process Instruments and Controls Handbook*, 2nd ed., (New York: McGraw-Hill Book Company, 1974).
9. Johnson, E. F., *Automatic Process Control*. (New York: McGraw-Hill Book Company, 1967).
10. Lloyd, S. G., and G. D. Anderson, *Industrial Process Control* (Marshalltown, IA: Fisher Controls Company, 1971).
11. Manning, A. W., and D. M. Dobs, "Design Handbook for Automation of Activated Sludge Wastewater Treatment Plants," EPA-600/8-80-028 (Cincinnati, OH: U.S. Environmental Protection Agency, 1980).

12. "Principles of Automatic Process Control," ISBN 87664-108-7 (Research Triangle Park, NC: Instrument Society of America, 1968).
13. Shinsky, F. M., *Process-Control Systems*, 2nd ed. (New York: McGraw-Hill Book Company, 1979).
14. Soule, L. M., "Automatic Control—A Reprint From *Chemical Engineering*," (Foxboro, MA: Foxboro Company, 1970.
15. Tucker, G. K., and D. M. Wills, "A Simplified Technique of Control System Engineering," (Fort Washington, PA: Honeywell, Inc., 1962).
16. Fertik, H. A., "Tuning Controllers for Noisy Processes," *ISA Transactions* 14(4):292–304 (1975).
17. Jury, F. D., "Fundamentals of Three-Mode Controllers," Technical Monograph 28 (Marshalltown, IA: Fisher Controls Company, 1973).
18. Wilson, H. S., and L. M. Zoss, "Control Theory Notebook." Reprinted from the *ISA Journal* (Spring House,PA: Moore Products Company).
19. Smith, C. L., *Digital Computer Process Control* (New York: Intext, 1972).
20. Harrison, T. V., " Minicomputers in Industrial Control," (Research Triangle Park, NC: Instrument Society of America, 1978).
21. Whitehouse, D., "How to Play It Safe With Analog Backup," *Instruments & Control Systems* (May, 1976).
22. Murray, J. J., "Combined Analog and Digital Control," *Chem. Eng.*, (June 21, 1976).
23. Harris, C. J., and Billings, S. A. "Self Tuning and Adaptive Control: Theory and Applications," Institution of Electrical Engineers (Stevenage UK: Peter Peregrinus, Ltd. 1981).
24. Baur, P. S., "Development Tools Aid in Fielding Expert Systems," *InTech* (April, 1987).
25. Bailey, S. J., "Artificial Intelligence in Industry: Expert Knowledge Bases in Control Loops," *Control Engineering* (December, 1986).
26. Finn, G. A., "Rules of Thumb for Implementing Expert Systems in Engineering," *InTech* (April, 1987).
27. "Environmental Conditions for Process Measurement and Control Systems: Airborne Contaminants," ISA-S71.04-1985 (Research Triangle Park, NC: Instrument Society of America, 1985).

Glossary

accuracy, measured: The maximum positive and negative deviation observed in testing a device under specified conditions and by a specified procedure. Usually expressed in terms of the measured variable, percent of span, or percent of actual output

actuator: A fluid-powered or electrically-powered device which supplies force and motion to a valve closure member

adaptive control: Control in which automatic means are used to change the type or influence (or both) of control parameters in such a way as to improve the performance of the control system

alarm: Device or function which signals the existence of an abnormal condition by means of an audible or visible discrete change, or both, intended to attract attention. (See also **annunciator**)

amplifier: A device that enables an input signal to control power from a source independent of the input signal and thus be capable of delivering an output that bears some relationship to, and is generally greater than, the input signal

analog: Pertaining to representation of numerical quantities by means of continuously variable physical characteristics. Contrast with digital

analog control: Implementation of automatic control loops with analog (pneumatic or electronic) equipment

analog signal: A continuously varying representation of a physical quantity, property, or condition such as pressure, flow, or temperature. The signal may be transmitted as pneumatic, mechanical, or electrical energy

analog-to-digital converter: A device used to convert an analog signal to approximate corresponding digital data

annunciator: A visual or audible signaling device and the associated circuits used for indication of alarm conditions

area control room: A location with heat and cooling facilities. Conditions are maintained within specified limits. Control rooms are commonly provided for operation of those parts of a control system for which operator surveillance on a continuing basis is required

automatic: Pertaining to a process or device that under specified conditions, functions without intervention by a human operator

backup: Provisions for an alternate means of operation in case of unavailability of the primary means of operation

baud rate: A unit of signaling speed indicating the number of signal changes per second. Most signal schemes have two states representing a bit equal to 1 or 0. In this case bit rate equals baud rate. Some signaling schemes have multiple states; in these, baud rate is less than bit rate

binary coded decimal (BCD):   Describing a decimal notation in which individual decimal digits are represented by a group of binary bits, e.g., in the 8–4–2–1 coded decimal notation, each decimal digit is represented by a group of four binary bits. The number twelve is represented as 0001 0010 for 1 and 2, respectively, whereas in binary notation it is represented as 1100

binary: A term applied to a signal or device that has only two discrete positions or states. When used for signals, the term denotes an "on-off" or "high-low" state

bit: 1. An abbreviation of binary digit; 2. A single character in a binary number; 3. A single pulse in a group of pulses; 4. A unit of information capacity of a storage device. The capacity in bits is the logarithm to the base two of the number of possible states of the device. Related to storage capacity; 5. One binary digit, the smallest piece of information in a computer system. A bit can be either 1 or 0

Boolean: An algebraic system formulated by George Boole for formal operations on true/false logic

bumpless transfer: A characteristic of a controller which permits control mode changes (such as automatic or manual selection) to be made without producing a discontinuity in the controller output

calibrate: To ascertain the outputs of a device corresponding to a series of values of the quantity which the device is to measure, receive, or transmit

cascade control: The use of two conventional feedback controllers in series such that two loops are formed, one within the other. The output of the controller in the outer loop modifies the set point of the controller in the inner loop

closed-loop: A signal path which includes a forward path, a feedback path, and a summing point, and forms a closed circuit

common mode rejection: The ability of a circuit to discriminate against a common mode voltage. Expressed as a dimensionless ratio, a scalar ratio, or in decibels

common mode voltage: A voltage relative to ground of the same polarity on both sides of a differential input

computer control system: All control action takes place within the control computer. Single or redundant computers may be used

control action: In a process controlling system, the output actions taken to achieve a desired result in the process

control center: An equipment structure, or group of structures, from which a process is measured, controlled, or monitored

control loop: A combination of two or more instruments or control functions arranged so that signals pass from one to another for the purpose of measurement or control of a process variable

control mode: A specific type of control action such as proportional, integral, or derivative

control sequence: See sequence control program

control system: 1. A system in which deliberate guidance or manipulation is used to achieve a prescribed value of a variable; 2. Refers to a system of hardware and software components including computers, disks, printers, instruments, control panels, operator facilities, communications channels, systems programs, development programs, and applications programs

controller: A device having an output that varies to regulate a controlled variable in a specified manner. A controller may be a self-contained analog or digital instrument or it may be the equivalent of such an instrument in a shared control system. A controller may be automatic or manual

data highway: 1. The physical media-connecting devices in a distributed control system, usually coaxial cable, fiber optic cable, or twisted, shielded pair; 2. The group of devices which provides communication among controllers and operator stations in a distributed network, including the cable, modems, processors, and the associated software to provide message handling, protocol, fault detection, time synchronization, and arbitration

deadband: 1. A specific range of values within which the incoming signal can be altered without also changing the outgoing response; 2. The range of values of a process variable where no control action is taken. If the process variable exceeds the deadband high or low limits, control action is started

deadman control: Continuous manual action (e.g., depressing a pushbutton) is required to modulate device position or speed. Device maintains status quo when control action is absent

dead time: The interval of time between initiation of an input change or stimulus and the start of the resulting observable response

derivative action: A controller mode which contributes an output proportional to the rate of change of the error

digital: Pertaining to representation of numerical quantities by discrete levels or digits conforming to a prescribed scale of notation

direct acting controller: A controller in which the value of the output signal increases as the value of the input (measured variable) increases

distributed control system: That class of instrumentation (input/output devices, control devices, and operator interface devices) which in addition to executing the stated control functions also permits transmission of control, measurement, and operating information to and from a single or a plurality of user-specifiable locations, connected by a communication link

disturbance: An undesired change that takes place in a process that tends to affect adversely the value of the controlled variable

error: The algebraic difference between the indication and the ideal value of the measured signal. It is the quantity which, algebraically subtracted from the indication, gives the ideal value

expendables: Items expected to be consumed such as print paper, lubrication fluids, and air filters. Distinguished from spare parts used for replacement of failed components such as printed circuit boards, power supplies, and fuses

failure: Loss of ability to perform a specified function

feedback: The return signal in a closed-loop system representing the condition of the controlled variable

feedback control: Control in which a measured variable is compared to its desired value to produce an actuating error signal which acts upon the process to reduce the magnitude of the error

feedforward control: Control in which information concerning one or more conditions that can disturb the controlled variable is converted, outside of any feedback loop, into corrective action to minimize deviations of the controlled variable

final control element: The device that directly controls the value of the manipulated variable of a control loop

gateway: 1. A device which interfaces one manufacturer's data highway system to other manufacturers' computers or controllers; 2. A protocol converter

hand/off/automatic (HOA):   Refers to a 3-position selector switch on a control panel. In AUTOMATIC a computer or logic in the panel controls the associated device. In HAND, the device is turned on from the local panel by the operator. In OFF, the device state is off

Hertz: Abbreviated Hz; a unit of frequency equal to one cycle per second

incremental control: Use of short pulses to increase or decrease the value of the controlled variable. Contrast with positional control

instrumentation: A collection of instruments or their application for the purpose of observation, measurement, recording, or control of physical properties and movements

integral action (reset):   Control action in which the output is proportional to the time integral of the input; i.e., the rate of change of output is proportional to the input

integral time: The time required after a step input is applied for the output of a proportional plus integral mode controller to change by an amount equal to the output due to proportional action alone

interlock: A mechanical or electrical device or wiring which is arranged in such a manner as to allow or prevent operation of equipment only in a prearranged sequence

lead-lag compensation: An electronic network or software used to influence the response of a control loop

linearity: Ability to achieve a straight-line response to an input signal

load shedding: starting and stopping of equipment to reduce electrical power demand

local area network (LAN): 1. For programmable logic controller systems, an industrial broadband network which uses a protocol structure defined by MAP; 2. For personal computers, a broadband or baseband network which uses a proprietary protocol

local control: Control operations performed either manually or automatically at a control panel located near the process or equipment

logging: Recording values of process variables for later use in trending, report compilation, or historical records

loop: See **control loop**

loop gain: The ratio of the change in the return signal to the change in its corresponding error signal at a specified frequency. Note: the gain of the loop elements is frequently measured by opening the loop, with appropriate termination. The gain so measured is often called the open loop gain

manipulated variable: The quantity or condition which is varied as a function of the actuating error signal so as to change the value of the directly controlled variable

manual control: Control operations are performed directly by a human operator. Two levels of manual control are possible; 1. local manual—the process is controlled manually from the local panel; 2. computer manual—the process is controlled manually through computer system operator display station

manual loading station: A manual electronic controller. Used here to refer to a controller whose output is adjusted manually by an operator from the front of a panel

manufacturing automation protocol (MAP): A networking protocol structure based on the International Standards Organization (ISO) Open Systems Interconnection (OSI) reference model. It is commonly used in factory automation to specify the connection of different manufacturers' devices into a network

monitoring: The information on the conditions of various processes, operations, and security obtained by electronic devices

normal mode voltage: A voltage induced across the input terminals of a device

offset: The steady-state deviation of the controlled variable from the set point caused by a change in load

on/off control: A system of regulation in which the manipulated variable has only two possible values, on and off

open-loop: A signal path without feedback

operating limits: High and low limits set for a process variable. A value of process variable between these limits is considered normal and no control action is taken. When either of the limits are exceeded an alarm or control action is initiated

optimization: A process whose object is to make one or more variables assume, in the best possible manner, the value best suited to the operation at hand, dependent on the values of certain other variables which may be either predetermined or sensed during the operation

optimizing control: Control that automatically seeks and maintains the most advantageous value of a specified variable, rather than maintaining it at one set value

package system: Equipment which is supplied as a system, including controls

panel: A structure that has a group of instruments mounted on it, houses the operator-process interface, and is chosen to have a unique designation

permissive: A signal which permits the placing of equipment into operation

piping and instrumentation drawing (P&ID):   A schematic drawing of the process showing all liquid flow paths, location of all sensors and instruments, and location of backup conventional control equipment

positional control: Use of an analog signal output value to modulate device position or speed. Contrasts with incremental control

power management: Making most efficient use of operational periods for equipment and using the lower cost rate, nonpeak electrical energy periods to the greatest extent possible. See **load shedding**

primary element: The device which converts a portion of the energy of the variable to be measured to a form suitable for amplification and retransmission by other devices. See **sensor**

process: 1. Physical or chemical change of matter or conservation of energy; 2. The collective functions performed in and by industrial equipment

process control: The regulation or manipulation of variables influencing the conduct of a process in such a way as to obtain a product of desired quality and quantity in an efficient manner. Descriptive of systems in which controls are used for automatic regulation of operations or processes

process variable: 1. In a control loop, the variable being controlled to the set point; 2. Any parameter within the process that is of interest from an operations or control standpoint

proportional action: A control action in which there is a continuous linear relation between the output and the input

proportional band: The change in input required to produce a full range change in output due to proportional control action

R/C filter: An electronic filter network made up of passive components such as resistors and capacitors

range: The region between the limits within which a quantity is measured, received, or transmitted, expressed by stating the lower and upper range values

rate time: For a linearly changing input to a proportional-plus-derivative mode controller, the time interval by which derivative action advances the effect of proportional action

ratio control: Control in which a secondary input to a process is regulated to maintain a preset ratio between the secondary input and an unregulated primary input

real time: The performance of a computation during the actual time that the related physical process transpires, in order that the results can be useful in guiding the process

regulatory control: Maintaining the outputs of a process as closely as possible to their respective set point values despite the influence of set point changes and disturbances

remote: Referring to an operations point located at some distance from the field device

remote/local (R/L):   Refers to a switch setting at local panel controls. The switch in local position means that control can be exercised at the local panel only, locking out control from a remote location. The switch in remote position means that control is exercised from a remote location and local controls are locked out

repeatability: The closeness of agreement among a number of consecutive measurements of the output for the same value of the input under the same operating conditions, approaching from the same direction, for full range traverses

repeats per minute: Controller integral mode adjustment units. The inverse of integral time

reset windup: In a controller containing integral action, the saturation of the controller output at a high or low limit due to integration of a sustained deviation of the controlled variable from the set point

reverse action controller: A controller in which the value of the output signal decreases as the value of the input (measured variable) increases

sensor (sensing element):   That part of a loop or instrument that first senses the value of a process variable, and that assumes a corresponding, predetermined, and intelligible state or output

set point: In a control loop, refers to the input variable that sets the desired value of the controlled (process) variable. It may be manually set, automatically set, or programmed

shared controller: A control device that contains a plurality of preprogrammed algorithms which are user-retrievable, configurable, and connectable, and allows user-defined control strategies or functions to be implemented. Control of multiple process variables can be implemented by sharing the capabilities of a single device of this kind

shared display: The operator interface device used to display signals and data on a time-shared basis. The signals and data, i.e., alphanumeric and graphic, reside in a data base from where selective accessibility for display is at the command of the user

signal: 1. The event or phenomenon that conveys data from one point to another; 2. A time-dependent value attached to a physical phenomenon and conveying data

simulation: The representation of certain features of the behavior of a physical or abstract system by the behavior of another system; for example, the representation of physical phenomena by means of operations performed by a computer or the representation of operations of a computer by those of another computer

software: 1. A set of programs, procedures, rules, and associated documentation concerned with the operation of a computer system; for example, compilers, library routines, and manuals; 2. A program package containing instructions for the computer hardware

span: The algebraic difference between the upper and lower range values

steady-state: A characteristic of a condition, such as value, rate, periodicity, or amplitude, exhibiting only negligible change over an arbitrary long period of time. It may describe a condition in which some characteristics are static, others dynamic

supervisory set point control system: 1. The generation of a set point or other control information by a computer control system for use by shared controllers, shared displays, or other regulatory control devices; 2. Control loops operating independently, subject to intermittent corrective action from an external source; 3. A high-level program whose primary function is to oversee an ongoing process and alter the general parameters of a control strategy based upon mathematical relationships

telemetering: The transmission of a measurement over long distances, usually by electromagnetic means

time constant: The time required for the output of a single capacity element to change 63.2% of the amount of total response when a step change is made in its input

transducer: An element or device which receives information in the form of one quantity and converts it to information in the form of the same or another quantity

transmitter: A transducer which responds to a measured variable by means of a sensing element, and converts it to a standardized transmission signal which is a function only of the measured variable

variable, directly controlled: In a control loop, the variable, the value of which is sensed to originate a feedback signal

variable, manipulated: A quantity or condition which is varied as a function of the actuating error signal so as to change the value of the directly controlled variable

variable, measured: A quantity, property, or condition which is measured

velocity limiting control: Control in which the rate of the change of a specified variable is prevented from exceeding a predetermined limit

Index